# TRANSPORT AND DEVELOPING COUNTRIES

Why do images of the Third World involve impassable roads, crowded railways, inefficient ports and congested cities – why are these associated with lack of development? These links are examined, first in theory and then in practice, for road, rail, air and water transport. Widely drawn examples illustrate good practice and bad and demonstrate there are no universal solutions to transport problems.

The introductory chapter examines the contrast between mobile and immobile societies, the links between transport and spatial organisation, the ways in which transport may influence development processes and the theories advanced to explain the relationship between the two. In the following chapters attention is directed to the individual modes, the form of their organisation and the ways in which they are able to influence the broader issues of spatial organisation and development.

The illustrative examples are selected to provide a balanced regional coverage. A concluding chapter identifies the main themes for the future in the evolving relationship between transport and development processes.

A pervasive theme of the book is the provision of transport services and facilities which are appropriate to defined development aspirations and the importance of always planning with particular conditions, locations and populations in mind. The book, therefore, will be of relevance to academics of all disciplines concerned with development problems and will also provide a wider context for those involved with the planning and development of transport projects of all kinds.

**David Hilling**, Member of the Chartered Institute of Transport, was formerly Lecturer in Geography at the University of Ghana and Senior Lecturer at Bedford College and Royal Holloway, University of London.

# TRANSPORT AND DEVELOPING COUNTRIES

*David Hilling*

London and New York

First Published 1996
by Routledge
2 Park Square, Milton Park, Abingdon, Oxon, OX14 4RN

Simultaneously published in the USA and Canada
by Routledge
270 Madison Ave, New York NY 10016

Transferred to Digital Printing 2006

Typeset in Garamond by
J&L Composition Ltd, Filey, North Yorkshire

*British Library Cataloguing in Publication Data*
A catalogue record for this book is available from the British Library

*Library of Congress Cataloging in Publication Data*
Hilling, David.
Transport and developing countries / David Hilling.
p.    cm.
Includes bibliographical references (p.    ) and index.
ISBN 0-415-13654-7. - ISBN 0-415-13655-5 (pbk.)
1. Transportation-Developing countries.   2. Developing countries—
Economic conditions.   I. Title.
HE 148.5.H55   1996
388'.09172'4-dc20       95-47892
CIP

Printed and bound by CPI Antony Rowe Ltd, Eastbourne

For all those overseas students it has been my pleasure to teach and from whom I have learned far more than they can possibly have learnt from me.

# CONTENTS

# FIGURES

viii

# PLATES

PLATES

# TABLES

# PREFACE

As originally conceived this book was to be similar in approach and effectively a 'sister' volume to my former colleague Alan B. Mountjoy's highly successful *Industrialisation and Developing Countries* (Hutchinson, 1963). With the passage of time this link has become less direct and possibly less useful but I hope that this volume will serve as a tribute to Alan Mountjoy, who was a pioneer in the area of geography and development and such a stimulus to those who had the good fortune to work with him. I cannot match his clarity or wisdom but I hope that this volume will provide some insights into the complex relationship between transport and development and the special problems of Developing Countries trying to provide transport systems which will serve their development aspirations and goals.

My background is that of transport geography and I hope that undergraduates doing advanced courses in this subject and also Master's degree students in development studies will find the book of relevance. However, I do not see the book simply as a text in transport geography but rather as concerned with the broader issues of transport and the development process.

The present literature on transport geography and development falls into three broad categories. First, there are a number of texts that deal in a systematic way with transport geography. Second, there are what might best be called monographs – specialist studies on individual modes or the transport of particular regions. Third, there are the regional geography and more general development texts which sometimes include a section on transport. My knowledge of Developing Countries and varied experience of teaching transport studies, academic and professional, in Britain and overseas, convinces me of the need for a book which attempts to combine elements of the three approaches. This book is a systematic, but necessarily selective, approach to transport geography from the point of view of the Developing Countries and with an emphasis on the role of transport in their development.

There is a vast and rapidly growing literature on the subject and in some ways this volume is a distillation from a wide variety of sources. It is an attempt, undoubtedly reflecting some personal bias, to bring together some of the theories and examples which are otherwise dispersed widely in the literature and not readily accessible to the many, from different academic or professional backgrounds, who may be interested in the subject. I am conscious that in this approach there is the danger that all of the readers will not be satisfied all of the time but I think that this is a risk worth taking.

The introductory chapter examines some of the main ideas regarding the complex relationship between transport and development and this general theme is also basic to the chapters on specific modes which follow. The approach adopted certainly reflects a personal interest in the technology of transport related as this is to a wide variety of geographical factors. As a geographer I am firmly wedded to the idea of the individuality of places, from which it follows that there can be no generalised, universal solutions to transport problems. Indeed, I would argue that all too many of the transport projects of Developing Countries have been based on an unthinking transfer of technology and strategies with too little regard for the context provided by people, place and time. In this I feel sure that the book contains messages of relevance for all those concerned with transport decision making.

I hope that the examination of the characteristics of the individual modes will make the book suitable for professional training courses and examinations, such as those of the Chartered Institute of Transport, which are of relevance to, and attract many students from, Developing Countries. For them I hope that the book will provide a broader context for their more detailed studies. The Chartered Institute is closely associated with TRANSAID which is specifically concerned with the problems of movement of aid and much of the content of this book has bearings on that subject.

In the preparation of this book I am greatly indebted to many people, not all of whom it will be possible to name. In many parts of the developing world there are transport managers and operatives and former students who have devoted time and effort to assist me in my travels and in data collection. I have special thanks for the library staff at the Chartered Institute of Transport and in particular Sue Woolley, Helen Berry and Françoise Mobbs. Justin Jacyno of the Geography Department, Royal Holloway, University of London, has provided invaluable assistance in the preparation of all the figures and and I am grateful to Sue May for photographic work. I am grateful to all those who provided photographs and in particular Cargill Plc, Barry Deacon (Utopia Fresh Produce, Zimbabwe), Pilatus Britten-Norman and the Yantian Port Authority. Through its direct involvement in the field and also through its publications, the Intermediate Technology Development Group has done more than any organisation to encourage a realistic approach to the transfer of technology to Developing Countries. I am especially grateful for their help in illustrating this book and I welcome this opportunity of drawing attention to their invaluable work.

My wife, Wendy, and son, Hugh, gave invaluable assistance in the final stages in the preparation of the manuscript. All errors of commission or omission remain mine.

David Hilling
Berkhamsted

# 1

# TRANSPORT AND
# DEVELOPMENT

Many factors contribute to economic and social progress, but
mobility is especially important because the ingredients of a
satisfactory life, from food and health to education and
employment, are generally available only if there is adequate
means of moving people, goods and ideas.

(Owen, 1987)

This means transport. Few would disagree with Owen's broad
conclusion regarding the relationship between transport and the
development process although the precise nature of the interaction
is far from clear. Much of the evidence from empirical sources is
contradictory (Howe and Richards, 1984) and this has spawned
considerable debate and a vast literature – 'there are few clear
statements of the ways in which the variable components of
transport improvement and economic advance are related' (Pawson, 1979).

## THE TRANSPORT GAP

It is easy to demonstrate that at the global level there are vast
differences in the availability of transport, indeed, that there is a
stark contrast between a relatively immobile Third World and the
highly mobile advanced economies (Owen, 1964, 1987). Large
parts of the Third World are characterised by lack of year-round
mechanised transport and movement is by unreliable, high-cost,
labour-intensive methods. Much of the infrastructure is poorly
maintained and in disrepair and is inadequate for present needs
without the complication of growth of demand in the future. The
skills and resources necessary to upgrade the transport are usually

1

lacking. The majority of the population live in spatially circumscribed local socio-economic systems and a relatively static state in which immobility and poverty are clearly related (Owen, 1987).

In the advanced economies recent years have seen a transport revolution with the advent of cheap, mass air travel, high and rapidly rising levels of personal mobility based on car ownership and containerisation of general freight with associated concepts of intermodalism, just-in-time delivery and restructured distribution channels. All of this has encouraged new levels of internationalisation in industry and commerce and for the favoured the world has shrunk to village scale. For the villager in the Third World little has changed.

The global discrepancies in transport provision are represented in Table 1.1 from which it can be seen that the 'developed' regions of North America, Europe and Oceania have 26.0 per cent of the earth's land area, 15.1 per cent of the population but account for 60.2 per cent of the world's commercial vehicles, 74.6 per cent of the cars and 53.6 per cent of the rail freight. In sharp contrast, Africa, Central and South America and Asia have 57.5 per cent of the area, 79.4 per cent of the population but only 39.4 per cent of the commercial vehicles, 20.4 per cent of the cars and 22.5 per cent of the rail freight. Central and South America has vehicles roughly in proportion to its population but Africa emerges overall as an area of serious under provision. Japan accounts for 64 and 67 per cent respectively of Asia's commercial and passenger vehicles and excluded from the totals leaves that area with a grave deficiency.

*Table 1.1*  The global transport gap, 1990 (% of world total)

|  | Area | Population | Commercial vehicles | Passenger vehicles | Rail freight |
|---|---|---|---|---|---|
| North America | 16.1 | 5.2 | 40.1 | 35.2 | 40.9 |
| Europe | 3.6 | 9.4 | 18.2 | 37.4 | 12.2 |
| Oceania | 6.3 | 0.5 | 1.9 | 2.0 | 0.5 |
| Former USSR | 16.4 | 5.7 | 0.2 | 4.8 | 23.9 |
| Central and South America | 15.1 | 8.5 | 8.7 | 6.7 | 3.7 |
| Africa | 22.2 | 12.1 | 3.5 | 2.0 | 3.1 |
| Asia | 20.2 | 58.8 | 27.4 | 11.7 | 15.7 |
| (excluding Japan) |  |  | (9.7) | (3.8) |  |

*Source:* United Nations Statistical Yearbook

Table 1.2 Vehicle stock growth rates (percentages)

| | Commercial vehicles | | | Passenger cars | | |
|---|---|---|---|---|---|---|
| | 1951–1971 | 1971–1990 | 1951–1990 | 1951–1971 | 1971–1990 | 1951–1990 |
| World | 350 | 230 | 750 | 370 | 220 | 790 |
| North America | 210 | 220 | 480 | 230 | 150 | 350 |
| Europe | 320 | 190 | 630 | 1,050 | 230 | 2,360 |
| Asia (excl. Japan) | 560 | 500 | 2,830 | 740 | 440 | 3,290 |
| Central and South America | 330 | 470 | 1,570 | 640 | 520 | 3,340 |
| Africa | 490 | 300 | 1,470 | 550 | 260 | 1,420 |

*Source:* United Nations Statistical Yearbooks

Between 1951 and 1990 (Table 1.2) the world total of commercial vehicles and cars grew by 750 and 790 per cent respectively, with rather slower growth in the second half of the period. Starting from a high base, growth in North America was well below the world average as was European growth in number of commercial vehicles. However, European car numbers expanded very rapidly, especially in the first two decades. Indeed, over the 40-year period the North American share of vehicles has declined, as has that of commercial vehicles in Europe. For all developing regions the growth in both categories has been far in excess of the world average and their share of the world total has also increased. Africa again emerges as the lagging region. Yet despite these growth rates, the gap between the mature economies and the Developing Countries is still vast and there is no prospect of this gap being narrowed significantly in the foreseeable future.

The gap is just as dramatic with respect to rail transport and while industrialisation in most of the developed countries was based on the expansion of railways there are many Developing

*Plate 1.1* Transport is an all-pervasive input in the development process and in many Third World countries is labour intensive and the infrastructure is unsophisticated and poorly developed (Photo: Intermediate Technology).

4

Countries, especially in Africa, which will never have a 'railway age' in their development profile (Chapter 3).

Owen (1964) devised an index of travel and freight mobility which he applied to a range of countries for the late 1950s and this he updated for his later work on mobility and development (Owen, 1987). A selection of the countries is shown in Table 1.3. The index of freight mobility was based on length of rail line per 100 sq km, rail lines per 1,000 population, surfaced highways per 100 sq km, ton-miles per capita and commercial vehicles per capita. The passenger mobility index was based on passenger miles per capita, cars per capita and railways and highways as for the freight index. In each case Owen related the indices to a base of 100 for France. Owen concluded that over the 20-year period there had been a widening of the gap between rich and poor although there had been some increases in freight movement in countries near the bottom of the scale. There had been little or no improvement with respect to passenger mobility and it was on the basis of these calculations that Owen distinguished between the contrasting worlds of the mobile rich and the immobile poor.

An index of transport which combines aspects of passenger and freight movement and for comparative purposes retains the base of 100 for France (Table 1.3) shows that the relative position of the immobile poor has continued to decline although there has been increased mobility in countries such as South Korea, Brazil and Thailand which have been moving into the Newly Industrialising category. The available statistics are incomplete, far from accurate and produce anomalies but nevertheless suggest a broad relationship between levels of income and mobility especially at the upper and lower ends of the range. The figures for the earlier years also suggest that freight movement was being given greater emphasis than passenger transport and it can be argued that this was ignoring a critical ingredient in the development process. However, in many Developing Countries the distinction between freight and passenger transport is far from clear and for them this blurring provides justification for merging the indices for passenger and freight mobility, as has been done in Figure 1.1.

That there is a link between levels of income and mobility would seem undeniable and it is but a short step to suggest that this is a causal relationship and that transport therefore has a special place in the development process. This was summed up in the oft-quoted contention that 'the material development of Africa may

Table 1.3 Indices of mobility for selected countries (base France 100)

| | 1957–8[a] | | | 1980[b] | | | 1990[c] | |
|---|---|---|---|---|---|---|---|---|
| | GNP per capita | Travel mobility | Freight mobility | GNP per capita | Travel mobility | Freight mobility | GNP per capita | Transport index |
| Canada | 150 | 149 | 223 | 95 | 114 | 374 | 133 | 136 |
| Sweden | 131 | 93 | 105 | 119 | 96 | 151 | 108 | 81 |
| France | 100 | 100 | 100 | 100 | 100 | 100 | 100 | 100 |
| West Germany | 105 | 99 | 91 | 117 | 101 | 57 | 105 | 70 |
| UK | 104 | 86 | 94 | 63 | 78 | 47 | 97 | 64 |
| Netherlands | 77 | 58 | 69 | 101 | 83 | 42 | 94 | 67 |
| South Korea | — | — | — | 15 | 8 | 16 | 46 | 31 |
| Egypt | 9 | 8 | 9 | 5 | 5 | 13 | 22 | 12 |
| Brazil | 14 | 20 | 20 | 18 | 18 | 23 | 18 | 25 |
| Mexico | 23 | 23 | 28 | 15 | 14 | 42 | 17 | 16 |

| | | | | | | | | |
|---|---|---|---|---|---|---|---|---|
| Venezuela | — | — | — | 31 | 24 | 36 | 17 | 26 |
| Chile | 32 | 36 | 38 | — | — | — | 15 | 21 |
| Argentina | 28 | 68 | 64 | 24 | 32 | 114 | 13 | 35 |
| Colombia | 21 | 9 | 11 | 11 | 6 | 47 | 10 | 14 |
| Thailand | 7 | 8 | 7 | — | — | — | 9 | 12 |
| Bolivia | 8 | 18 | 20 | — | — | — | 7 | 12 |
| India | 6 | 12 | 10 | 2 | 5 | 26 | 2 | 9 |
| Ghana | 15 | 10 | 10 | — | — | — | 3 | 3 |
| Ethiopia | 3 | 2 | 2 | 1 | 2 | 3 | 2 | 1 |
| Bangladesh | — | — | — | 1 | 2 | 3 | 2 | 2 |

*Notes:* a  From Owen, 1964
b  From Owen, 1987
c  Based on surfaced roads per sq.km., vehicles per 1,000 population, rail travel per person per year, rail tonne-km freight per 1,000 population, domestic air km per person per year

*Source:* Based on United Nations Statistical Yearbooks

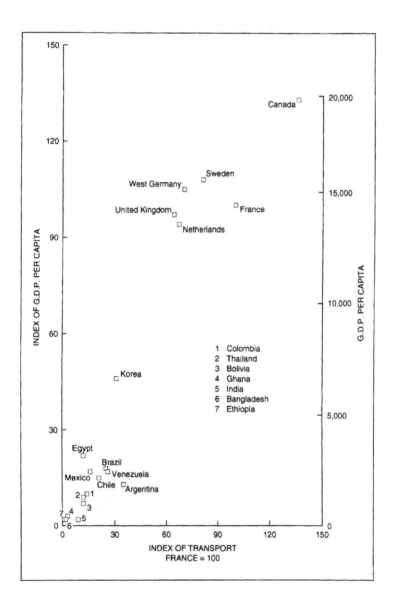

*Figure 1.1* Indices of transport
*Source*: Calculated from UN Statistical Year Book (1990, 1991)

be summed up in one word – transport' (Lugard, 1922). Leaving aside doubts about the emphasis on material development, this simplistic interpretation of the role of transport undoubtedly found expression, post-1945, in international and national development strategies. For the United Nations transport provided 'the formative power of economic growth and the differentiating process' (Voigt, 1967) and a lack of transport was seen as 'a major deterrent to rapid economic growth and social progress' (United Nations, 1967).

In response, some 40 per cent of World Bank loans in the 1960s went to transport projects and in many Developing Countries in the 1950s and 1960s transport investment formed an even higher proportion of fixed investment (Bejakovic, 1970). Having designated the years 1968 to 1978 as World Development Decade, and possibly dissatisfied with the outcome, the United Nations then embarked on the Transport Decade (1978–1988). In 1994 transport was again selected for special treatment by the world body. There is almost daily evidence of the inadequacies of transport systems to deal with the special conditions created by drought, famine and political conflict but also a persistent feeling that a root cause of these conditions is lack of development and therefore the need for better transport. There is clearly a real danger that the argument becomes circular. At the heart of the debate is the question of which comes first – development or transport? To this chicken–egg argument it has been suggested that there is only one honest answer – 'we do not know but probably both come first' (Lowe and Moryadis, 1975).

A more useful response may well be that the transport–development relationship is both time- and place-specific. There can be no universally applicable model but rather a range of local variants deriving from particular conditions of geography, history, politics, economics and culture.

## DEMAND- AND SUPPLY-LED MODELS

To suggest that transport leads to development is to adopt a supply-led approach and there is a school of thought which argues that at early stages in the development process, and by implication in many Developing Countries, it is the provision of transport which leads to widening of markets, increased production and associated multiplier effects of an economic and social nature

(White and Senior, 1983). This is the traditional view of transport although for many development planners transport became just one element in a range of 'infrastructure' which also included energy, health, housing and education – expenditure on such essentials being classed as Social Overhead Capital. This capital was invested in anticipation of the development which it was thought would follow.

There are certainly many cases where development could be seen to follow the provision of transport – the new railway lines into the then Gold Coast (1903) and Northern Nigeria (1912) stimulated spectacular expansion of cultivation for export – cocoa in the case of the former and groundnuts in the latter (Mabogunje, 1980). Leinbach (1975) demonstrated the significance of transport for the stimulation of rubber cultivation and tin mining in Malaya while the opening up of Amazonia, for better and for worse, followed the construction of access roads into isolated forest areas and associated expansion of agriculture, forestry and mining (Kleinpenning, 1971).

An alternative view is that transport provision is invariably a response to demand and it is rarely developed except where there is a demand. Indeed, it can be argued that even in the supposedly supply-led situations cited above there was also a crucial and possibly more important demand factor operating. Demand for transport is of several distinct forms. Most obviously there is the *revealed demand* as expressed in the journeys that are actually made or the goods that are transported – the use that is being made of the existing infrastructure. However, at any place and at any point in time there is likely to be an element of *latent demand* – this comprises components of the existing demand which cannot be satisfied perhaps because of inadequacies in the infrastructure or prohibitive cost and which might be called delayed demand, and also completely new demand that may be created by additional or improved infrastructure. To revert to the previous examples, development followed transport because there was a latent demand – there was a market for the produce if the transport gap between producer and consumer could be closed.

What in fact is in question here is the existence or otherwise of what a number of writers have called prior-dynamism (Wilson, 1973; Leinbach, 1975). Pawson (1979), in a similar vein, refers to the set of conditions creating a need for transport innovation. These conditions could include the availability of readily exploita-

10

ble natural resources, the existence of entrepreneurs and the absence of restrictions of a political or economic kind (Wilson, 1973) – overall, an environment in which economic opportunity can be exploited. In other words, transport will have its most obvious impact on development where there is some latent demand or prior dynamism.

It is for these reasons that the supply-led models of transport must now be viewed with some caution. The cyclical innovation theory of Schumpeter (1934) in which expansion was seen to follow great innovations (e.g. the steam engine) which were the prime movers in the growth process and Rostow's stages of growth theory (Rostow, 1960) in which railways were seen as the most powerful initiators of economic take-off, clearly elevated transport to a special, even unique, role in the development process. There can be no doubting the powerful impact of railway building on the development process in many countries (see Chapter 3) and many have argued for a causal linkage (Jefferson, 1928; Rostow, 1960). However, it has also been argued that the cost of rail transport in nineteenth century North America gave it little advantage over other modes and it was not therefore indispensable (Fogel, 1962) and that the development of the railways followed rather than led to growth in other sectors. Fishlow (1965) in contrast argued that the railway was the critical element in the economic transformation of North America.

As defined above there can in fact be few places where there is not some element of latent demand and where all demand is already satisfied. However, what is not clear is whether the existence of such demand is in itself an adequate reason for transport to be developed. We are possibly in error in viewing transport simply in terms of demand and economic activity. Long ago it was pointed out that 'there can be no adequate theory of transport which has regard only to . . . the economic aspect . . . one must examine severally its relations to various social institutions – military, political, economic, ideal' (Cooley, 1894). Certainly in the case of some of the early African examples already noted the railways were built to penetrate hinterlands and so demonstrate the effective control required by the Congress of Berlin as a condition for laying claim to colonial territory and in a country such as Brazil improved transport has been a means of consolidating the political control by central government (Dickinson, 1983).

While transport provided in response to non-economic stimuli

11

may well have been successful on its own terms the broader impact on economic development does in all probability depend on the existence of latent demand. Given the parsimony of most colonial regimes it was vital that even politically motivated transport developments should pay their way (Mabogunje, 1980) and not be a burden on the colonial government and there was invariably one eye on the potential traffic and revenue earning capacity. This was certainly true of Ghana's railway and it was a considerable element of latent demand which led to the rapid and profound economic impact of many of the early African railways.

In contrast, it is also the case that some of Africa's recent railway construction has been conspicuous for its lack of impact – Nigeria's Bauchi extension of the 1960s and the northern extension of the Cameroon railway in the 1970s provide obvious examples. It seems clear that transport of itself will not necessarily result in economic development, or as Owen (1964) has argued, it is a necessary but not sufficient factor. It has been claimed that this is equally true of many of the other ingredients in the development process (capital, technical ability, education, natural resources) and that it is not therefore a very helpful idea (Wilson, 1973). However, it is helpful in pointing to the idea that development is based on a range of inputs and is most likely to take place where these are considered not separately but as part of a package and planned in a coherent way. This is certainly how transport should be viewed because it is rarely an end in itself and the demand derives from other sectors of economic and social life.

Gauthier (1970) distinguishes three possible relationships between transport and development. The first he terms *positive* – where an innovation in transport is demonstrably responsible in a direct way for expansion of economic activity. The second is the *permissive* effect where transport does not itself stimulate economic growth but is such that it does not inhibit such growth when other stimuli are operating. The third relationship may be classed as *negative* – the situation in which the returns on investment in transport are less than from the same investment in directly productive activity with the possibility of an actual decline in per capita income. Transport projects can be very demanding of capital and with borrowed money, high interest rates and only very slow or even negligible generation of income they can very easily have a negative impact. It is of course for this reason that Developing Countries can ill afford mistakes in the provision of transport infrastructure.

Adler (1971) has argued along similar lines in classifying transport projects. There are cases where the transport is an integral part of a development project and in which the emphasis is on the whole project and the adoption of least-cost transport solutions. The development of the LAMCO railway in Liberia for the evacuation of iron ore to the exporting port provides an example (Chapter 3). There are also situations in which all other requirements are satisfied and only suitable transport is lacking – this is rather a case of prior dynamism. Finally, Adler identifies those situations in which transport is built into an area of promise and where other investments will have to take place if development is to follow. Arguably, such 'act of faith' transport provision will rarely be justified unless other investments are made.

A related idea is that all productive activity requires some infrastructure, including transport, before it can become operational and this either exists as surplus capacity or it must be provided. Development by surplus clearly equates with the supply-led situation described above and if there is some surplus capacity there is always the possibility that 'footloose' investments in economic activity can be accommodated. Arguably, it is always wise to ensure that there is just enough surplus capacity to provide for reasonable anticipated growth. In contrast, development by shortage depends on a build up in demand with the possibility that infrastructure is not provided until demand justifies it but directly productive activity is not established until the necessary infrastructure is available – another of the vicious circles which bedevil life in Developing Countries (Hilling, 1978). Pawson (1979) probably comes as close as anyone to providing a satisfactory explanation of the transport–development relationship. His starting point is a set of conditions which create the need for transport innovation. There then comes the adoption of a transport innovation which over time is likely to diffuse spatially. Without going into the reasons, this diffusion process is likely to be very uneven spatially – it is easy to find examples of this in canal, railway and surfaced road networks and technologies like containerisation. There will then be the impact of the transport in stimulating both forward and backward linkages. The forward linkages are the developments which result from transport innovation (e.g. farmers changing to cash crop production alongside a new or improved road) while the backward linkages are the multipliers resulting from the transport (e.g. the impact of railway construction on the demand for iron and steel, coal and engineering industries).

It was these backward linkages which persuaded Rostow (1960) of the leading role of railways in the take-off to sustained growth in countries such as the United States, Britain and Russia. However, even in these countries the railways never consumed more than about 20 per cent of the iron produced (Pawson, 1979) and it is now generally accepted that the Rostow stages of growth model is a far from satisfactory basis for explaining the progress of development in the Third World. The conditions of nineteenth century Britain are not replicated in Developing Countries – most obviously they may have to import everything that they need to build a railway system and this greatly reduces, although does not necessarily eliminate, the possibility of backward linkages. India is one of the few Developing Countries in which railways have had a considerable backward linkage effect, and is now able to sell its expertise and technology to other countries.

In practice there are few Developing Countries with railway systems of a size to justify the development of substantial backward linkages even if they have the technology and resources to do so. It is worth noting at this point that in contrast road transport has been more successful in this respect and in many Developing Countries there is a considerable 'shade-tree mechanics' industry and a number of countries have moved into vehicle assembly with varying degrees of local production of components of the more basic type and in suitable environments have started rubber cultivation to feed rubber and tyre producing factories – Ghana has done this.

Given the global trading situation it is not now easy for Developing Countries to create backward linkages of the type and on the scale which followed railway construction in what are now the developed countries although the possibilities in activities related to road transport hold out more promise. The main hope must rest with the forward linkages and these in effect constitute what has been identified as latent demand.

## TRANSPORT AND FORWARD LINKAGES

It is relevant in this context to ask what is meant by development? For all too many, development has been synonymous with *economic growth* and measurable in terms of indices such as per capita income. Growth may simply mean more of the same and says nothing about change in structures whether economic, social or political. Many others have concentrated on *modernisation*, an idea espoused particularly by geographers who could map the diffusion of modernising behaviour – often emphasising material possessions such as telephones, televisions, radios and cars or the use of banks, schools, cinemas, medical or educational services (Leinbach, 1976). Some saw this as an undesirably Eurocentric approach to development questions and argued that development should be concerned with equity and *distributive justice* at all scales (Mabogunje, 1980). This clearly involves the idea of regional planning to bring about more even benefits from development. Some go further to argue that what is needed is a *socio-economic transformation* of the whole mode of production, an approach favoured by Marxist theorists, and involving the breakdown of colonialism, dependency and capitalism. This places a new emphasis on human issues, full mobilisation of society and the redefinition of a country's external relations. The idea of under-development turns the

*Plate 1.2* Until Band Aid money made possible a temporary bridge, an inadequate ferry across the Chari river provided land-locked Chad with its principal route for external trade and drought relief aid.

15

focus on subjugation to, and dependency on, external forces, removal of independence and loss of self-reliance (Mabogunje, 1980).

It does not take much effort to see that behind all these ideas about development there are considerations of distribution, movement, contact, diffusion and spatial interaction, in a word, transport. While in detail the complex relationship between transport

*Table 1.4*  Transport and development

| *Market consequences* | *Extra-market consequences* |
|---|---|
| 1  For users of transport services | |
| Vehicle size, character | Tourism |
| Transport operating costs | Recreational, amenity |
| Cost of time | Improved safety |
| Financial position of transport firm | Integration |
| Reliability, speed of transport | Improved information |
| Commodities carried | |
| Freight flows – volumes, direction | |
| Passenger flows – numbers, direction | |
| Changed distribution channels (handling, warehousing, inventory) | |
| Price changes for commodities | |
| | |
| 2  For non-users in zone of influence of transport facility | |
| Changes in cost of public services | Impact on general well-being |
| Changes in value of land for all uses | of community, region |
| Changes in value of crops and natural resources | Sequent occupance – extensive/intensive |
| Changes in rural land use | Emergence of entrepreneurial |
| Changes in urban land use | capacity |
| | |
| 3  Wider regional/national impact | |
| New patterns of investment | Changing patterns of internal/ |
| Changes in employment opportunity | external links |
| Changing patterns of income distribution and level | Changing relative significance of settlements, regions, |
| Changes in balance of trade, terms of trade | sectors |
| Spread of money economy | Demographic changes – structure, migration |
| Changing patterns of public finance – taxation, revenue | Changing investment criteria |
| | Changing political alliances, structures |

*Source:* Modified from Stokes (1968)

and the development process may not be fully understood it is not difficult to identify some of the ways in which transport may have an influence on productive and social activities.

Any new, extended or improved transport infrastructure will affect the range, capacity and the cost of movement, effectively providing positive changes in mobility and accessibility which will potentially enhance economic and social opportunity. There is no guarantee that local communities will respond to such opportunities (Wilson, 1973) and the empirical evidence suggests very considerable variations in the manner and extent to which they do so. Reactions are not readily predicted and the economic consequences allow for little more than speculation (Howe, 1984).

The variety of responses to transport improvement can be appreciated in terms of the range of possible consequences, many of them interrelated but few of them inevitable. An attempt is made in Table 1.4 to classify some of the main consequences of transport improvement as they affect the users of the transport, the non-users in the zone of influence of the transport facility and the wider regional or national impact (Stokes, 1968). The compartmentalisation implied in the table does not reflect reality in which distinctions may be far from clear but it does draw attention to a variety of non-economic considerations which some feel have been given too little attention in many studies of transport and development (Leinbach, 1976) and which are implicit in the definitions of development considered above.

## CONSEQUENCES FOR USERS

Many of the consequences of transport improvement derive from the reduced cost of transport services. The first beneficiaries will be the transport operators themselves, who, with better roads or rail track are able to use larger units with reduced costs. The cost benefits are likely to be greater with the first major development (e.g. new surfaced road) and rather less with each successive minor improvement. This explains why the impact of similar transport improvements in developing and developed economies is generally far more obvious in the former. A study of agricultural production in Argentina concluded that road improvement did not have the impact of a new road (Miller, 1973) and a study in Sabah suggested that production was not very sensitive to small reductions in cost (Bonney, 1964). This could well reflect the fact that transport costs

17

are often only a relatively small proportion of the final cost of many products. Also, one cannot be sure that reduced costs to the operator will be passed on to the shipper and the trickle down effect to the farmer might be slight or possibly take a long time to have any effect. Howe (1984) has argued that even where vehicle operating costs are reduced substantially the impact on, say, crop production, is not likely to be great in conditions in which either the farming or the transport services are arbitrarily regulated or under monopoly control rather than subject to market forces.

In cost-benefit analysis of transport projects in mature economies the value of time can often be a critical factor yet there is often the implication that this is less important in Developing Countries both in contrast to developed countries and also relative to other factors involved. While it is certainly difficult to calculate (low wage values, non-productive use of lost time) there are certainly situations in which an attempt at evaluation could usefully be made (Howe, 1984).

The reliability and speed of transport is critical for a range of goods and becomes more important as one moves from basic, bulk commodities to consumer products which may be perishable, have short shelf life or have restricted market windows. Howe (1984) gives examples of improved roads in Swaziland and St Lucia which reduced damage to citrus and bananas, respectively, and brought benefit to producers. The enhanced capacity for air freight in recent years has opened up a wide variety of new long distance trades in perishable products from Developing Countries to European and North American markets – for example, fruit, vegetables, flowers and sea food (see Chapter 4).

Changes in the range of commodities carried is therefore a normal consequence of transport improvement often accompanied by an increase in the volume. It may also be that the directional balance of movement is altered and while better transport allows local people to move their produce to markets it also enables goods from outside to be brought in more easily – this could be both to the advantage and possibly the disadvantage of the local economy and a factor in the process of under-development.

As a general rule the ability to move goods more quickly and with the assurance that delivery time will be short allows for a reduction in stock carrying and lower inventory costs. A possible consequence will be centralisation of warehousing facilities. With components air freighted from France to its assembly plant in Kaduna, Northern Nigeria, the Peugeot car company was assured daily deliveries and while an eventual change to container transport by sea and rail reduced transport costs it provided a less reliable and vastly longer delivery time for components.

A number of writers (Blair, 1978; Hay, 1973b) have pointed to the significance of roads for the movement of people and while in many Developing Countries it is impossible to distinguish clearly between passenger and freight transport (many vehicles are adaptable to variable combinations of the two (e.g. Ghana's 'mammy wagons') it does often appear that road transport is dominated by the former. A possible change of attitude on the part of operators is evidenced by reports from Zimbabwe of bus drivers showing increasing reluctance to take farmers' produce, especially of a bulky kind (J.A. Smith, 1989). In time, and as quantities to be marketed by individual farmers increase, there may well be the need for a sharper distinction in provision for passengers and freight movement or farmers will acquire small commercial vehicles for their own use.

Transport improvements can produce a rapid response in passenger numbers. Many of these people will be carrying trade goods because personal marketing of produce is normal in many Developing Countries. A study in Sabah (Bonney, 1964) suggested that in smallholder economies the facility with which people can be moved is of greater importance than movement of freight. Others will be travelling to medical centres, schools or to visit friends and relatives (Jones, 1984). In many African countries where kinship bonds are strong it is customary to return 'home' for religious and family festivals and improved transport facilitates and encourages this.

Almost inevitably there will be a time lag before economic activity responds to transport improvement and this is related to production cycles in agriculture and the often lengthy lead time for establishing new agricultural or industrial enterprises. On a priori grounds one might therefore expect 'pure' freight transport to respond more slowly than passenger numbers and any survey undertaken soon after transport improvement has been made may well not show any marked increase in freight movement.

With respect to the Rigo road improvement in Papua New Guinea there was noted a considerable increase in vehicle ownership and growth in all forms of traffic with new commodity movements and commuting patterns (Ward, 1970). In this case, and it could be fairly typical, the growth of road traffic was high initially and then slowed down. It is quite normal for diffusion following an innovation to conform to a flattened 'S' form – gradual growth at first, then accelerated and finally a slowing down as possibilities for response are exhausted. This has been well illustrated by the spread of containerisation.

In addition to the consequences in the transport market *per se* there are a number of extra-market effects. These could include tourist and recreational travel and the social travel already noted, and there will be cost savings with respect to the time devoted to such travel and for all users the cost benefit of improved safety. Far more difficult to quantify in monetary terms are the benefits to users of being able to communicate more effectively, of having more information available and being better integrated into political, social and cultural processes. It could certainly be argued that these are at the heart of development in any wider sense of the term. Wilbanks (1972) noted that the improved information flow resulting from better access contributed more to agricultural development than reduced transport costs.

## CONSEQUENCES FOR NON-USERS AND WIDER REGION

Development is not simply concerned with those who in a direct way use the transport facilities provided – of great importance will be the wider consequences in the zone of influence and these are likely to show decay with distance away from the facility itself. Some of the consequences in this category represent the more obvious geographical manifestations of the development process in action – the changing economy, landscape and welfare of the people living adjacent to the new or improved transport infrastructure. Indeed, much of the research into the impact of transport on development has been concerned with attempts to measure just such changes and Howe (1984) has provided a valuable review of the literature as far as rural roads are concerned, although many of the works cited provide insights into the wider transport– development relationship.

The reduced cost of transport implicit in improvements to infrastructure should be reflected in the cost of many goods and also the cost of providing a range of public services which will benefit not only the direct users of the transport but also the non-users. This is probably a main factor in the opening up of new land but is of considerable significance in extending the amenities and services which are such an important part of the overall well being of the people involved – educational, health and agricultural extension services come immediately to mind. Such social benefits should not be under estimated.

The value of land is a function of the uses to which that land can be put and improved access and lower transport costs can have a profound effect. Access to wider regional or international markets has been a main factor stimulating the change from subsistence to cash cropping with resultant changes in land values. In Zimbabwe a rural roads programme resulted in great expansion of farmer participation in marketing and big increases in cash crop production both of staple food crops (the country has even become a net exporter) and industrial cash crops such as cotton (J.A. Smith, 1989). While there are some exceptions, most of the evidence points to increases in cash crop production consequent upon road improvement with lower-value subsistence crops replaced by higher-value market crops. The extent of this change would certainly appear to be related to distance from the transport facility and also possibly the distance from main centres of population, commerce and external market links.

With intensification of production, land rents and values will rise and in areas of communal land holding but personal usufruct, as in much of Africa, there will be growing competition for land which will increasingly enter into the commercial market often with mounting litigation over issues of ownership. Changes in land use and land values consequent upon transport improvement are certainly not restricted to rural, agricultural areas although in many Developing Countries this is likely to be the most obvious impact.

To the extent that transport increases the utility of resources which may have no value to the community in whose area they are located it is able to increase the value of a whole range of natural resources. The iron ore deposits of Mount Nimba in remote northern Liberia only assumed 'value' on the opening of the railway and port which allowed mining and export to commence. The same would be true for the forest resources along the line of rail and some of the roads constructed in the east of Liberia. In the Amazonas area of Brazil there has been an all too rapid increase in forest clearance both for timber production and also to make way for new agriculture. In the same region mineral exploitation has been associated with road construction (Kleinpenning, 1971, 1978).

A detailed questionnaire examination of the consequences of rural road improvement in selected areas of peninsular Malaysia (Leinbach, 1976) suggests a very wide range of impacts and variety of persons affected. There were indications of rising land values, expanded range of buying and selling opportunities for local and imported produce, growth of small businesses, formation of growers' cooperatives and the creation of new economic opportunities (logging, vending). The improved roads stimulated inward and outward migration, increased access to medical services, increased acceptance of family planning, establishment of new postal services and a general increase in the availability and exchange of information. Journey making and transport ownership has been encouraged. While Leinbach is cautious regarding the precise relationship between the road improvements and the changes described there can be no doubting its significance.

It can be suggested that what is initiated for the wider region is a process of 'sequent occupation' in which there is likely to be a shift along the continuum from self-sufficiency to dependency, from a non-money to a money economy, from isolation to incorporation, from extensive to intensive activity and from rural to urban. There will be increasing scope for those with entrepreneurial ability to capitalise on their skills.

By extension from the above the wider region may well see new patterns of investment and the spread of the money economy which with increased employment opportunities will enlarge the base for revenue collection for local and central government. With time the whole pattern of trade could change with respect to the commodities involved, the markets and relative monetary values. There is considerable evidence to indicate changing patterns in income levels and distribution. The whole pattern of the region's or nation's external links could change, as they did with railway penetration routes of the colonial era, and internally there are likely to be changes in the relative significance of specific locations and settlements, regions and economic sectors.

Transport can at the same time stimulate both dispersion and concentration of people and activities. Settlements at favoured transport nodes will attract population and services. Improved transport links draw people to them and bring in-migration but

22

*Plate 1.3* In many Developing Countries there is no clear distinction between passenger and freight transport and vehicles are chosen for their flexibility of function.

by the same token make it easier for population to migrate out-wards – the net effect being different from place to place.

## 'DISBENEFITS' AND COSTS

It would be all too easy to conclude from the above that transport improvement in some ill-defined way is able to stimulate a chain of benefits at all levels. Indeed, the emphasis on transport in much earlier development planning was based on this very assumption and the definition of development merely in simple economic growth terms. In some cases even economic growth did not take place yet the wider reality is that the effects of transport improve-ments are often very different from those envisaged and in all too many cases undesirable or harmful in terms of the process of change initiated and the impact on equity and social justice in the community concerned. Transport can have a negative impact in a far wider sense than defined economically by Gauthier (1970) or Adler (1971).

As Pirie (1982) has suggested with respect to the railways in Southern Africa while they did bring capitalist development and promoted growth in some localities they also had some debilitating effects and acted as agents of underdevelopment as well as development. Particularly, this was reflected in greater indebtedness, greater dependency, the promotion of wage labour and proletarianism and the facilitation of migrant labour and the social problems this created. In different degrees much of this is true for most transport developments.

Some of these negative impacts will now be considered but one can but agree with Brookfield (1975) when he argues that it is impossible to separate the positive from the negative consequences of change and evaluation must be based on the defined goals in each situation. At the very broadest level it can be argued that the transport improvement of the colonial era was fundamental to the underdevelopment of the colonial territories resulting in regional imbalance, core–periphery relations, chains of exploitation and dependency (McCall, 1977). The interests of the local population were often secondary to the economic advantage of the mother country and there was the tendency for capitalism to be reinforced (Edwards, 1978). The colonial penetration produced export orientation and the emergence of the colonial economy characterised by a small range of primary exports and heavy dependence on imported consumer goods and, increasingly, foodstuffs. A possible effect of this was the stifling of traditional economic activities. This was certainly true for the 'cottage' industries in many parts of Africa and India (Dickenson, 1978) and food crop farming in areas in Africa.

A similar effect of transport improvement, albeit different in scale, may be identified at the national or local level. Transport improvement is, by definition, highly selective spatially, and its impact is bound to be spatially differentiating – the higher the technological level the more marked this will be. Motorway style roads and high-capacity rail links make up but a small fraction of the networks of which they are a part and numerous studies have pointed to the concentration of economic activity and benefit along narrow corridors or in 'islands' favoured by the higher capacity facilities (Taaffe *et al.*, 1963; Hoyle, 1983, 1988). In the hierarchical development of transport networks there is implicit a spatial inequality – those near the facility benefit more than those in the surrounding area, those in the city more than those in town and village.

It is possible to go on from this to suggest that improved transport may be a factor in social stratification, widening income differentials and greater inequality (Elmendorf and Merrill, 1977) with centralisation reducing the possibility of spreading benefits (Howe, 1984). There is evidence to suggest that increased cash cropping leads to increased income inequality, differential increases in land values, sharpened land competition, concentration of land in hands of relatively more prosperous members of the community and possibly increased unemployment. While overall income may rise the distribution may be worsened and poorer farmers may even be forced out. It has been suggested that in some cases, especially in remoter areas, new roads may have a harmful impact on minority groups, some of the Indian groups of Amazonia provide the obvious examples, and possibly even lead to their elimination. In contrast, as Howe (1984) has noted, where a minority group has control of local economic activity, such as the Lebanese or Indians in parts of Africa or the Chinese in parts of South East Asia, they may benefit disproportionately. In Nepal, Blaikie *et al.*, (1977) noted that new opportunities consequent upon road improvement tended to be taken up by the already advantaged and these were often 'outsiders'.

Where the existing transport involves labour-intensive technology the effect of improvement may well be reduced employment in the transport sector itself and resultant increases in hardship. In rural areas the elimination of head porterage could have this effect as could the change from pedal rickshaws to modern transit in urban areas (see Chapter 7). Likewise, a change from labour-intensive to mechanical construction and maintenance methods, especially for roads but also to some extent for other transport facilities, will mean reduced employment and demand for skills that are not available locally and may have to be imported at great cost. It is open to debate whether or not it makes sense in situations of abundant cheap labour and high levels of unemployment to change to methods which consume scarce capital which could possibly be used to better purpose in other areas of development.

It is certainly the case that transport improvement has a profound impact on the distribution of population and the migratory forces that shape this distribution (McMaster, 1970) and in Southern Africa allowed the widening of labour markets so that an area such as southern Mozambique became little more than compounds for workers in the Rand (Pirie, 1982). The whole process of rural–urban migration, a main problem in many Developing Countries, is undoubtedly facilitated, although not necessarily caused, by improved transport.

It must also be noted that both in its construction and operational phases transport can have serious implications for the environment (Farrington, 1992), to which must be added the environmental degradation which may result from the economic activities initiated by or associated with the transport infrastructure. The environmental impact may be local in scale such as drainage impeded by a transport route, soil creep and land slips on bare slopes, but can be global as in the case of large scale forest clearance in Amazonia, tropical Africa and parts of Asia.

## TRANSPORT AND SPATIAL ORGANISATION

The idea that transport is fundamental to spatial organisation is developed in many standard texts and in a variety of ways (Lowe and Moryadis, 1975; Taaffe and Gauthier, 1973). One of the best-known theories of transport (Ullman, 1974) revolves around the idea that geographical reality (climate, resource endowment) results in natural *complementarity* between places and that this will result in interaction or exchange if problems of transfer can be overcome. This *transferability* will be a function of the availability and cost of transport. This is not greatly different from the classical economist's concept of comparative advantage and the emergence of trading patterns based on division of labour and specialisation at varying levels from local through regional to global. The idea that such a spatial specialisation would ultimately lead to an equalisation of incomes has been shown not to be realistic and growing disparity is the all too frequent outcome.

This concept of transferability is closely associated with the idea of *mobility*, the ability of people and goods to move or be moved or the capacity to change place (*Oxford English Dictionary*) and it is facility of movement to particular places which determines their *accessibility*. Mobility and accessibility are at the heart of any consideration of the organisation of space and the means by which ends in other sectors are accomplished (Owen, 1987).

Mabogunje (1980) has suggested that the process of development is in effect one of spatial reorganisation – new social goals require new types of spatial action and transport is the means by which these actions are integrated. Many of the models of transport development (Taaffe *et al.*, 1963; Hoyle, 1973; Rimmer, 1977) are effectively designed to identify phases in the evolution of spatial organisation. While earlier models such as those of Taaffe may not deal effectively with issues of modal interlinkage and spatial integration and say little about important questions of control and therefore power (Mabogunje, 1980), more recent models have gone much further in attempting to incorporate transport into a wider political, administrative and even social context (Rimmer, 1977).

In Rimmer's model the starting point is a *pre-contact phase* in which there are no overseas, external links and there is an unmodified traditional system of production and circulation whose character is determined by the culture of the people and resources of the area concerned. Movement will be by low capacity tracks and possibly waterways. In the *early colonial phase* there were the first contacts by sea between the more and less advanced territories, establishment of trade and the gradual spread of this trading influence inland from the coastal points of contact. Yet for the most part the Europeans themselves did not penetrate inland, their coastal footholds were tenuous and usually by courtesy of the coastal inhabitants (Hilling, 1969a) and there was minimal adaptation in the traditional societies.

This was to change with the gradual emergence of what might be called *full colonialism* when the core territories consolidated their dominance over the periphery. This was reflected in the building of roads, railways and ports, establishment of political control, new administrative settlements and the diversification of economic activity. There was considerable spatial restructuring as new settlements were created, as mining and plantation agriculture were established, as rural–urban migration was initiated and as Europeans, in varying numbers and degree of permanence depending on the environment, set up their homes. There emerged a structural and behavioural dualism (Brookfield, 1975) which in much simplified terms may be considered as traditional and modern. In terms of economic activity and transport the dualism was well defined with the externally imposed transport developed for the export oriented 'modern' sector and integrated into the global economy, and built with scant regard for, and having little in common with, that which served the traditional sector based on local production, marketing, technology and circulation.

Political independence for the colonies was not always followed by the economic independence which the Developing Countries might have wished for and this was reflected in a phase of *neo-colonialism* with the continuation of the established core–periphery dependency relations, albeit in a more subtle and possibly more invidious form. However, in the Developing Countries more positive approaches to development planning brought a call for improved infrastructure and transport services better adapted to demand. In Sierra Leone the old colonial railway was eventually abandoned and many Developing Countries embarked on rural road schemes (Chapter 5). It has been claimed with some justification that in practice there have been few radical changes to inherited transport networks (Hoyle and Smith, 1992) and there is still an all too sharp distinction, one might say a gulf, between the transport that has been externally imposed and that which is needed at the local level. This is a point which will be developed at a number of places in later chapters of this book.

It has been suggested that a *mature independence* situation can be identified (Hoyle and Smith, 1992) in which longer periods of political independence lead to meaningful domestic decision making, a more varied range of external links and the intensification of the domestic infrastructure. Australia and Canada could be cited as examples but it is not easy to identify Developing Countries that have as yet reached this stage – when they become 'developed'.

## TRANSPORT TECHNOLOGY AND DEVELOPMENT

For the most part this discussion of transport and development has been couched purposely in very general terms. In practice, the impact of transport on wider development issues will be a function of a number of interrelated considerations and where transport does not produce the anticipated development it may be that it is the transport itself which is at fault. For example, *timing* is critical and inappropriate timing in relation to demand or the provision of associated services and inputs to other sectors may be counter productive. Likewise, the *location* of each new transport facility in relation to the overall geography of demand is important and this could range from macro-scale decisions regarding the siting of new airports or seaports or the location of new rail links and motor-ways, to the micro-scale detail of road or rail alignment and location of small wharves on rivers.

It is also critical that the decision makers are appropriate and that decisions regarding transport are made for the right reasons – there have certainly been cases in many Developing Countries where decisions have been made by politicians for wholly personal gain or the perceived advantage of their home region irrespective of the real merits or likely development impact of the transport facility concerned.

Possibly of even greater significance will be questions related to the technology selected for the transport provided and consider-able emphasis will be given to this in the modal chapters which follow. If transport is to have any impact on development it must be appropriate, in the fullest sense of that term, to the desired development goals. It must be the most suitable mode and level of technology and in the right place at the right time – all too often transport developments have failed to satisfy these conditions and it is hardly surprising that the resultant impact has not measured up to expectations.

Table 1.5 represents a hypothetical evolutionary scheme for an agricultural development project (Deschamps, 1970). At each stage the character of the demand can be identified and appropriate transport adopted. At the *feasibility stage* the demand will be mainly from small groups of 'expert' surveyors (hydrologists, pedologists, agronomists, anthropologists, etc.), probably from overseas aid agencies or consultancy firms, who will need to move around 'in the bush' and off-track. This demand is likely to be met by the use of four-wheel-drive, rugged vehicles or in certain conditions by helicopters or boats. There is unlikely to be any real track con-struction although there may be some ground clearance.

Having produced a scheme acceptable to government it is possible that the *extension services* will be involved in education and demonstration to encourage local acceptance and participa-tion.This could involve larger numbers of personnel, possibly with local knowledge and understanding, travelling widely, visiting vil-lages and farmers and possibly setting up trial schemes. This may require a larger range of vehicles, could involve the use of some 'traditional' transport technology and could be constrained by issues of vehicle availability, operational costs and limited bud-gets. It is unlikely that track construction will be necessary

*Table 1.5* Agricultural development and transport demand

| | Stage | Transport demand | Transport technology |
|---|---|---|---|
| 1 | *Feasibility*<br>Field surveys | Small numbers of persons<br>Some equipment<br>Off track | Helicopter, boats, rugged<br>four-wheel drive<br>vehicles |
| 2 | *Extension*<br>Education<br>Demonstration | Larger numbers of staff<br>Mobility at local level<br>Access to farms, villages | Horse, cycle, moped,<br>   smaller four-wheel<br>drive<br>   vehicles |
| 3 | *Input*<br>Implementation | Movement of imported<br>   equipment, machinery<br>Movement of workers<br>One-off demand | Improved port facilities<br>Improved access roads<br>Heavy lorries, low loaders<br>Rail link (?)<br>Buses (?) |
| 4 | *Output* | Movement of produce<br>   (volume, nature of<br>   commodity, regularity,<br>   seasonality)<br>Export markets<br>Domestic markets | Improved regional road<br>   links<br>Specialised vehicles<br>   (tankers, refrigerated)<br>Rail transport<br>Water transport<br>Upgraded port facilities<br>Terminal, storage<br>   facilities |
| 5 | *Income* | New patterns of<br>   consumer demand –<br>   imports, domestic<br>   production<br>New levels of personal<br>   mobility – education,<br>   health, social, work<br>Local re-investment –<br>   new industries,<br>   expanded agriculture<br>Settlement growth<br>Expanded services and<br>   amenities | Upgraded distribution<br>   channels<br>Warehouse, transit<br>facilities<br>Specialised vehicles<br>Sophisticated port<br>facilities Air transport<br>expansion Higher<br>capacity roads<br>Rising car ownership<br>Increased demand for<br>   public transport |

*Source:* Modified from Deschamps (1970); Hilling (1976)

although some improvements may be needed to allow movement of larger vehicles or possibly give year-round access.

In what may usefully be called the *input phase* more radical innovation may be needed. Depending on the precise location of the project area, access may have to be greatly improved. There may be the need for earth-moving equipment, pumps, generators, irrigation pipes and construction materials, all of which may have to be imported, thereby creating a demand for improved port facilities and regional link routes of all-weather capability and higher capacity than already exists. The construction of these new routes and facilities will themselves require equipment and access. Special vehicles such as low loaders or cement transporters may be needed and fuel will be required in larger quantities. However, it is worth noting that many of the inputs and the movements that they generate are one-off in character and the track and vehicles may be redundant on completion of the project works. The relative merits of the different modes will certainly be an issue to be carefully considered in relation to the nature of the movements generated and anticipated future growth of traffic. Thus, railways are likely to have a far higher 'import' element than roads and are not suitable for one-off movements of 'out-of-gauge' equipment. However, they may well be thought suitable if the expected output of the operational project is to be regular movement of large volumes of produce on a year-round basis.

This brings one logically to the *output phase* when the scheme starts producing the commodities for which it was planned. These may be for domestic or export markets, the volume and character of the goods and the production cycle over time will be different for each commodity, and the required transport and handling facilities may well be specific to each. Short-life perishables (flowers from Kenya, seafood from Côte d'Ivoire) will require transport with very different characteristics from bananas or pineapples or large volumes in bulk such as cocoa, coffee, palm produce and latex. Speed may be the critical requirement or specialised vehicles (refrigerated, tanker, side-loading, top-loading, container) may be necessary. Decisions will have to be made regarding mode and suitable technology and higher capacity track, whether road, rail or water, may have to be provided to accommodate larger, regular flows of traffic.

The outcome of all this new production, hopefully, will be *income* both personal and public, and new patterns of trips will be generated (educational, medical, social) and new forms of consumption demand created (foods, cars, household goods) which may have to be met by imports or more desirably by new or expanded domestic industries. Either way there is likely to be demand for port facilities of greater capacity or sophistication (Chapter 7) and a possible increase in international air transport

*Plate 1.4* A cotton ginnery opened at Lalago, Tanzania, in 1994, is dependent on road transport over poor roads but reduces long hauls for the farmers and is a boost to local cotton production and income (Photo: Cargill plc).

(Chapter 4). There may well be a demand for high capacity inter-regional links. The impact of these changes may well be reflected in urbanisation and the need to upgrade urban transport systems (Chapter 6) and there is likely to be a growing disparity between the transport provision in the core and peripheral regions.

While this example is concerned with an agricultural project the underlying logic would be just as applicable in the context of an industrial scheme, and more importantly, it illustrates a vital principle that has been ignored in much of the theorising about and also the actual planning of transport. It is not enough to think only of transport in general terms but essential that careful consideration is given to the *when* (timing), *where* (location), and *what* (technology) in each particular case. In effect these must be related to the overall development objectives – this is a demand-led approach but always with realistic anticipation of future growth and likely changes. It is an emphasis on transport that is appropriate for each situation and set of conditions.

## THE ELEMENTS OF TRANSPORT

In the more detailed examination of transport which will follow certain main elements should be borne in mind. All transport involves a *way*, *terminals*, *vehicles* and *energy*. These vary in significance between modes, between different levels of technology, between places and with time.

The *way* may be natural (sea, air), improved natural (dredged river channel) or artificial (road, railway, canal). The construction and maintenance costs depend on the type of way and range from zero in the case of oceanic routes to massive for high technology roads and railways. The extent to which the costs are passed on to the user depends on whether the way is common user, as in the case of most roads, or single user as for most railways.

In a broad sense a *terminal* is any point at which there is access to a way or at which interchange between modes is possible – what might be called an intermodal interchange. Terminal facilities vary greatly in size, design and level of sophistication as between modes and also for any particular mode – there is a world of difference between, say, a simple fishing port and Rotterdam, but both effectively perform transfer functions between the land and sea modes. Often very complex in design (large airports or seaports), the terminal may be multi-functional and in addition to the obvious transfer function may provide storage, processing and servicing facilities.

The *vehicle* is the unit of carriage and can vary greatly in capacity, speed, degree of specialisation, flexibility and propulsion and ranges from simple carrying aids for human porters, through the variety of road and rail vehicles to large planes and ships. The movement of any vehicle will consume *energy* in some form and this is the most obvious way in which the cost of transport reflects the effort of overcoming the friction of the earth's surface.

In combination, the way, terminals and vehicles determine both the *flexibility* of the overall transport system and also its *capacity*. In any consideration of the role of transport in stimulating spatial interaction and development these are vital questions for the transport planners. As we shall see in later chapters, road and rail transport differ greatly in their flexibility and capacity but so do different types of road and different designs of railway.

It has been suggested (Lowe and Moryadas, 1975) that geographic space consists of points (nodes) at each of which different types of activity take place and as human needs can rarely be

33

satisfied at any one point there is need for interaction between points. Our ability to effect this interaction will be in part a function of distance but also a reflection of the transport that is available. Transport networks evolve through a process of node connecting which may in theory, but not always in practice, involve links between points at successively greater distances apart (Lachene, 1965; Taaffe *et al.*, 1963; Haggett and Chorley, 1969). To the extent that a transport link is provided with surplus capacity it serves not only to provide for immediate flow between the nodes but also for longer-term relative growth of these nodes – what has been called the accumulator effect to the network (Haggett and Chorley, 1969).

Because the resultant networks can be reduced to a series of points (nodes) and lines (links) they become amenable to geometric and mathematical analysis and this provides a basis for the comparison of different networks. Haggett and Chorley (1969) provide a valuable synthesis of previous work and ideas and a basis for much more recent work.

Several ideas of particular relevance to the present discussion may be noted. The shortest, straight line link may well be the ideal (Hay, 1973a) but it may be desirable to deviate from this for positive reasons (e.g. route length is increased to touch more points and generate more traffic and revenue or possibly for social reasons) or it may be necessary to deviate for negative reasons (e.g. to negotiate difficult terrain). The differing sensitivity of road and rail transport to gradient (see Chapters 3 and 5) provides obvious reasons for deviation. Deviation, for whatever reason, will be reflected in cost of construction and maintenance but this may be offset by a greater impact on traffic generating activity along the route.

Basic to the relationship between transport and development, however the latter is interpreted, are the associated concepts of connectivity and accessibility. *Connectivity* within networks can be compared on the basis of the number of links per node (the Beta index), the ratio between the actual and maximum possible number of links the (Gamma index), the number of loops in the network (the Cyclomatic number) and the ratio between the actual number and maximum possible number of loops (the Alpha index). In simple illustration the indices for the basic Ghanaian rail network at two points in time are shown in Figure 1.2. In similar calculations rail systems invariably produce a lower order of connectivity than roads, a vital consideration for development, and Developing

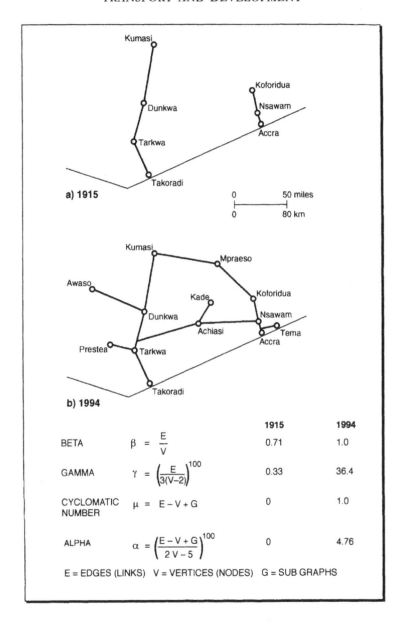

*Figure 1.2* Ghanaian railway network – indices of connectivity

Countries usually have far lower levels of connectivity for all modes than developed countries (Kansky, 1963). A detailed study of a large number of countries (Garrison and Marble, 1962) showed a close correlation between high indices of connectivity and high technical and socio-economic status.

*Accessibility* may be interpreted in two ways. At the one level it is related to connectivity and may be defined as the number of links that have to be traversed, or sub-journeys made, from any particular node to reach all other nodes, in effect the 'reachability' of that node (Lowe and Moryadas, 1975). At another level, accessibility may mean the number and distribution of places at which one can join the network. For example, it is possible to walk, or even perhaps drive, on and off an unengineered path in open country at virtually any point whereas there are but few points at which one can join a motorway. Likewise, roads of all types apart from motorways have a fineness of access that is not possessed by railways. The implications with respect to spatial impact are all too obvious and this is basic to the idea of differential settlement growth and development described by McMaster (1970). We shall return to these ideas in discussing the different modes.

Other characteristics of networks will also have implications in relation to spatial interaction and development. The route length in relation to surface area, the *network density*, has been shown to be closely related to population density and networks frequently assume a *geographic orientation* which favours movement along certain axes and in some directions rather than others. Networks also tend to develop *hierarchic structures* and these clearly relate to similar differential growth in settlements and the services they provide and the hinterlands they serve – this is basic to central place theory. Networks can also be said to have differing levels of *efficiency* which in addition to those characteristics already noted would include a variety of technical features related to surface, gradient, curvature, alignment and capacity.

It is usually the case that networks have within them a variety of weak points. These may be the result of inadequacies in connectivity, accessibility or the efficiency with which the system operates and where, for whatever reason, the capacity of the network is inadequate in relation to the demands placed upon it there may be evidence of *stress*. This may be reflected in congestion, delay, accidents, loss of time in transit, deterioration of the system and increased cost. While few if any transport networks are devoid of stress points it is probably the case that in many Developing Countries they are the norm rather than the exception and examples will be identified in later sections of this book. An essential aspect of transport planning is the identification of potential stress points well in advance so that action can be taken.

With these general considerations in mind attention will now be directed to the specific modes of transport with particular emphasis on their technical characteristics, organisation and relationship to the broader issues of spatial organisation and development outlined above.

# 2

# INLAND WATERWAYS
# TRANSPORT – THE
# 'NATURAL' MODE

In many respects inland waterway transport has been the neglected mode and does not usually feature prominently in the transport planning strategies of most countries (Hilling, 1980; Deplaix, 1989; Tolofari, 1984). Perhaps this is because waterway transport is the most 'natural' and least obtrusive mode and much of it takes place, especially in Developing Countries, by low-technology means which are widely dispersed, not easily regulated or controlled and for which the collection of tolls and statistics is difficult or even impossible. Yet at times of increasing concern for the environment and the long-term need, but not the immediate will, to economise on fossil fuels, water transport recommends itself as the mode which should be encouraged to the full. For the Developing Country there is the additional advantage that where there are waterways their transport capacity can often be increased substantially at low cost in comparison with other modes. This will be of particular significance where there is the potential for moving large volumes of bulk goods.

It must always have been the case that wherever people could see the possibility of floating objects, water was used for transport. The methods adopted to achieve this end were as ingenious as they were varied (Hornell, 1946) and not surprisingly reflected local availability of materials. Aids to floating, whether of people or goods, have included logs, wooden blocks, gourds, inflated skins, bundles of reeds and pottery jars. On Lake Bosumptwe in Ghana, people 'ride' wooden floats and Tamil fishermen did likewise in former times. The Zak people of the Upper Indus connected inflated goat skins on a wooden frame and a similar technology was found on the Oxus, in upper Egypt and on some Chinese rivers. Gourds were used on Lake Chad, in northern Nigeria, the

Blue Nile and lower Egypt. Reeds boats came in all shapes and sizes in places as far apart as Lake Titicaca, the Nile and Asia.

Both logs and reeds could be bound together and shaped to make craft of considerable size (Heyerdahl, 1971) with dry well decks for the safer carriage of goods and people. In India and the peninsula and archipelago areas of South-East Asia and also in Africa, dug-out canoes exist in great variety and where suitable timber was available (Malabar Coast, British Columbia, New Zealand) could be of considerable size. Balsa wood logs were used in South America (Heyerdahl, 1950).

While this infinite variety of craft is part of the history of water transport it is also part of the present – all of the types mentioned and many others are in active use today and represent just one end of a spectrum of technology at the other end of which are the high technology push–tow systems of the Mississippi, Rhine, Yangtze and, as time goes on, an increasing number of waterways in Developing Countries. The distinct advantage of water transport is this range of technology, unit capacity and ability to respond to and reflect local conditions of environment and demand.

Water transport provided the first form of bulk transport and it is not perhaps surprising that the first urban cultures were to be found in the riverine lands of Mesopotamia, Egypt, North-West Indian subcontinent and China – the so-called hydraulic civilisations. In these areas water transport provided a capacity and range not matched by human porterage or animal transport and allowed movement of surplus agricultural production and building materials on which the new urban populations were dependent. Water transport assisted in the development of non-agricultural population and the specialisation which is a vital component in economic development, and facilitated administrative, spatial integrations and contact between different cultural groups. Many of the world's largest cities grew up on navigable rivers – Calcutta, Bangkok, Chongquing and Cairo to name but a few.

There is not always a clear dividing line between inland, coastal and maritime transport and on large rivers, estuaries and in near-coastal zones the distinctions may become very blurred. The inland and coastal transport may well be undertaken by the same craft, for example some of the smaller junks of China (Maitland, 1981), and historically there were certainly cases where journeys started as inland or coastal and by accident became maritime. Some of the cultural diffusion in the Pacific region was almost certainly of this kind. In the planning of water transport it would certainly be unwise for too rigid distinctions to be enforced.

## THE WATERWAYS

Navigability is a function of the relationship between the physical characteristics of the waterway and the type of craft used. Navigable waterways are of many forms and while rivers may be the most obvious it is well not to forget other types. Coastal waters sheltered by off-shore islands or reefs, coastal lagoons such as those of Benin or Côte d'Ivoire in West Africa or the 'backwaters' of India's Kerala coast and the distributaries of the great deltas (Nile, Niger, Ganges) all provide for navigation.

Lakes are also important. Lake Titicaca, 160 km in length and straddling the Peru–Bolivia border is, at 3,800 m, the world's highest navigable waterway. It has well-developed port facilities such as the Peruvian rail terminal at Puno and the Bolivian port of Guaqui and a number of passenger and freight vessels which have been imported in parts and assembled at the lakeside, one of which is of 850 tonnes capacity. Lakes Malawi, Tanganyika and Victoria, together with some of the smaller African lakes, provide for local transport and in some cases are parts of important international through routes. On Lake Chad the craft are mainly traditional in style with a few larger motorised launches and on the lakes of Kashmir small launches are to be found.

Water courses constructed for drainage or irrigation, as in Pakistan, Egypt and Sudan, may be used for transport as may the lakes resulting from hydro-electricity schemes such as Kariba and Cabora Bassa on the Zambesi, Kainji on the Niger, Ghana's Lake Volta, Itaipu on the Parana and Bassa Bonita on the Tiete river. In some cases (e.g. dams on the Parana, Kainji on the Niger) the dams have been provided with bypass locks to allow for through navigation. It is a significant but often forgotten advantage of water transport that the track may be multi-purpose (irrigation, drainage, flood control, water supply, transport of water) and it may be possible to spread the investment costs amongst several end users. It is rarely possible to do this with other forms of transport.

River navigation varies greatly from one river to another and on any one river is likely to change over the length of its course and also possibly from season to season (Lederer, 1979). It follows that there may be variability of capacity and range for inland waterway transport such as would not normally be found with railways or better engineered roads. To some extent this variability of navigation can be reduced by engineering works and navigation aids but can rarely be completely eliminated as experience on the Rhine and Mississippi shows only too clearly. The cost of total elimination would often be prohibitive and not justified by the benefits.

The completely artificial canal specifically for navigation is a common feature in Britain, mainland Europe and parts of north-eastern United States but is not a common feature in most Developing Countries. However, China is a country with a long tradition of waterway construction dating back to the Bian Canal in Henen of the fourth century BC, the 'Magic' canal of Guangzi of 219 BC and the most famous Jing Hang (Grand Canal) from Hangzhou to Beijing of the period AD 581–617 (Hadfield, 1986). By the eleventh century the Chinese had developed the pound lock. In the 1980s the Grand Canal was straightened and deepened in sections over 400 km to take barge tows of up to 2,000 tonnes capacity. Few other transport arteries are likely to compete with this canal's life span of over 13 centuries.

In Guyana, the Dutch built canals to assist sugar cane production and these also served as transport arteries but the nineteeth century plans for a canal link in the Buenos Aires area and another in the Rio de Janeiro province of Brazil came to nothing. There have been several attempts to canalise the Digue river link from Carthagena to the Magdalena river in Colombia and a 1950s rebuilding still carries some traffic. In Sri Lanka, the Dutch built a series of canals to link river and lagoons, the largest of which was the 132 km Colombo to Puttalam canal, and this was in use until road-transport competition and rising maintenance costs proved too much in the 1950s.

## RIVER NAVIGATION

River navigation, as distinct from that on canals, is a function of a variety of interrelated geographical factors such as river flow, river long profile, channel cross-section and channel alignment. In the case of long rivers it may well be the conditions in the remote source regions which determine water levels and navigability in the middle and lower reaches.

In many Developing Countries the *river flow* and water level

*Plate 2.1*   Like many tropical rivers, the Niger, here near Timbuctoo, has seasonal variations in flow which may restrict navigation for larger vessels for parts of the year.

reflect the marked seasonality of rainfall which characterises many tropical regions (Lederer, 1979). In the case of the Nile, which receives 78 per cent of its water from the Blue Nile and Atbara rivers with sources in the summer rainfall Ethiopian Highlands, the ratio between mean and maximum flow is 1:13. The Zaire river, which has tributaries from both north and south of the equator, has a more even flow and the mean–maximum ratio is 1:3. Rainfall in the Guinea Highlands in summer produces the main flood in the Niger river but this takes about five months to work downstream and is 'flattened' by its progress through the inland delta from Segou to Timbuctoo and then added to by the flood from the Benue river. The mean–maximum discharge ratio is about 1:3 at Niamey and 1:6 at Lokoja further downstream. On the upper Benue tributary the navigation window is a mere six to eight weeks from August to October for barges of 1.2 to 2.1 m draught. There are similar flow variations on most tropical Asian and South American rivers and each river will have a unique flow regime.

However, even these seasonal patterns of flow can vary greatly from year to year. The Niger was greatly reduced in flow during the period of low rainfall in the mid-1980s. To complicate matters, superimposed on these broad patterns are localised events such as particularly heavy storms (e.g. hurricanes and typhoons in coastal regions) which can alter river levels , sometimes dramatically, in a very short time. Hadfield (1986) describes the Ahrewady (Irrawaddy) on which 'navigation is, however, hindered by varying water levels, and sometimes made hazardous, by sudden storms that bring high winds, or cloudbursts that cause water surges sideways from dried-up water course'. The normal difference in water level between low water (December–April) and high water resulting from the Himalayan snow melt and monsoon rains (May–October) can be as much as 12 m. While such variations are not restricted to tropical rivers – the flow of the Rhine or Mississippi, for example, can vary considerably – it is reasonable to conclude that water levels for navigation are far less predictable over time in tropical environments. This clearly creates problems for the development of water transport.

The *long profile* of any river will reflect the geological conditions, tectonic history and erosion processes in the region. The Amazon and its tributaries have generally low gradients over very long distances and there are few interruptions to navigations in the form of rapids or falls until the upper courses are reached. As a result, even sea-going ships of moderate size can navigate the Amazon over 3,700 km to Iquitos and shallow draught vessels a further 780 km to Achual Point. On the 2,000 km Orinoco, navigation is possible over 1,100 km to the Atures rapids.

In sharp contrast are most African rivers, where there is characteristically an alternation of low-gradient, braided sections and higher-gradient sections of cataracts, gorges or falls as rivers cut through the breaks of slope between plateau surfaces at different levels. All too often there is one of these interruptions relatively close to the coast and this was a main impediment to European exploration of the continental interior. The Zaire river falls 300 m in its last 400 km and navigation inland is restricted to the 150 km to Matadi, the head of navigation for sea-going ships. Under the 1870s Leopold plan a railway was constructed to bypass the falls between Matadi and Stanley Pool on the lower river and also around breaks in navigation between Kisangani and Ubundu and above Kindu. On the Nile the first cataract is 800 km upstream

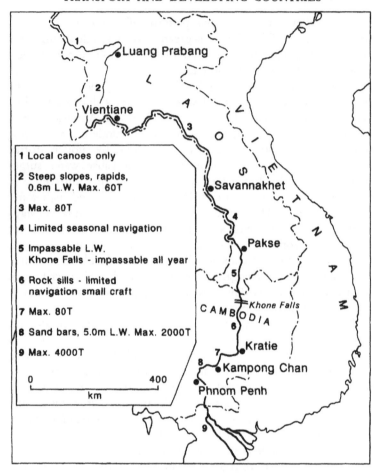

*Figure 2.1* Navigation on the Mekong River
*Source:* Modified from Somabha (1979)

from the delta but this is still very much in the lower course of a river that is 6,500 km in length. The Niger has a similar pattern of interruptions to navigation.

The Mekong river is comparable with those in Africa and its long profile has numerous points at which navigation is restricted or even impossible, the longest navigable section being some 600 km from Savannakhet to Vientiane (Figure 2.1).

The river *channel cross-section* is another factor influencing navigability and will depend on gradient, flow and processes of erosion and deposition. It follows that channel characteristics are continuously changing. While water depths may change in deeply incised rivers, in gorges for example, it is often in the areas of lower gradient (middle Niger, Ganges valley, Ahrewady) that river channels change most dramatically and often very rapidly to create problems for navigation. This is a characteristic that is particularly marked in tropical rivers which carry very heavy sediment loads in suspension. The mean sediment flow of the Rio Magdalena in Colombia is seven tonnes a second and the resulting 31.5 million tonnes of silt a year causes great channel instability in the lower reaches (Danish Hydraulics, 1988). The annual sediment load of the Hwang Ho is 1,600 million tonnes (Walling, 1981) and dredging to maintain navigation channels can be a problem in all such rivers.

The changeability of such rivers can be illustrated by the Parana river. In 1983 heavy flooding led to serious siltation and dredging failed to restore channels to their original depths. Just four years later in 1987 low rainfall resulted in further reduction of river levels and the stranding of a number of vessels (*Lloyd's List*, 7/10/1987). The Forcados river channel in the Niger delta silted from 6.0 m in 1899 to 4.7 m in 1936 at which time it was decided to gain access to delta ports by way of the Benin river. This too silted up and after 1940 the Escravos river mouth became the main access channel but a depth of 6.4 m is now maintained with difficulty, even after the construction of training moles in the 1960s.

So variable are the tropical river regimes that for planning purposes even a single parameter such as least available depth of water, the critical factor for navigation and craft size, can usually be expressed realistically only in terms of probabilities. This is illustrated for the Parana–Paraguay rivers in Figure 2.2. (Sas *et al.*, 1985).

Related to the channel cross-section and long profile is the *channel alignment*. However, it is important to distinguish between the degree of meandering of the river itself and, of greater significance for navigation, the meandering of the deep water channel within the river. Both vary over time and the channel alignments can change greatly in the shortest time scales. The channel alignment and the nature of the bends are critical determinants of the length of craft that can be utilised on any stretch of the river – this will be true for self-propelled craft on smaller

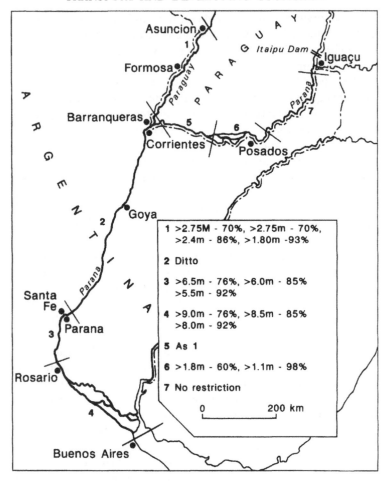

*Figure 2.2* Draught probabilities on the Parana-Paraguay rivers
*Source:* Modified from Sas *et al.* (1985)

waterways and especially for integrated multiple barge 'tows' whether these are pushed or pulled on larger rivers. The geometry of channel bends will also affect the ability of units to pass one another and on European waterways a minimum curve radius of 10 times the vessel length has been adopted. On many tropical rivers, the Ganges provides a good example, tow size has to be restricted. For artificial canals there is a similar design relationship between alignment and permissible craft size.

It is possible to assist larger craft around tight bends with the use of a bow thruster which provides lateral thrust. Such a thruster unit may be built into the bow of self-propelled craft or may be in the form of a detachable, hydrodynamic bow boat with an engine which can be fixed at the head of a string of pushed barges. Such units are in common use on American and some European waterways and have recently been introduced on Chinese waterways.

Overall, the critical natural parameters for navigation will be depth of navigable channel, width of channel, river flow velocity, minimum bend radius, minimum width in narrow sections and the ratio of the wetted cross-section of the waterway to the mid-ship section of the ship. This ratio influences water resistance, power needed, speed, manoeuvrability, the effect of the vessel on channel banks and bed and the tendency for the vessel to 'squat', thereby increasing its draught. The relationship between these various factors is shown in Table 2.1. As speed increases the channel depth and wetted cross-sections need to be increased. The European Class IV waterways (barges of 1,350 tonnes capacity) adopt a channel width of 28.0 m (excluding lay-by areas), minimum water depth of 3.5 m over a width of 28.0 m, an 'n' coefficient of 7 and

*Table 2.1* Navigation parameters for a single line of barges

| | | | | *LASH* barge | | *SEABEE* barge | | *BACAT* barge |
|---|---|---|---|---|---|---|---|---|
| Capacity (tonnes deadweight) | | | | 340 | | 840 | | 147 |
| Length (m) | | | | 18.7 | | 29.7 | | 16.8 |
| Beam (m) | | | | 9.5 | | 10.7 | | 4.7 |
| Draught (m) | | | | 2.5 | | 2.7 | | 2.4 |
| Immersed section (sq. m) | | | | 23.2 | | 29.2 | | 11.4 |
| Desired channel depth (m) | | | | 3.7 | | 4.2 | | 3.7 |

| *Barge train speed (kph)* | *Coeff. (n)* | *Channel wetted section (sq. m)* | *Surface/ bottom width (m)* | *Channel wetted section (sq. m)* | *Surface/ bottom width (m)* | *Channel wetted section (sq. m)* | *Surface/ bottom width (m)* |
|---|---|---|---|---|---|---|---|
| 9 | 7 | 162 | 47.5/40.1 | 204 | 52.8/46.4 | 80 | 25.3/17.9 |
| 11 | 10 | 232 | 66.4/59.0 | 292 | 73.7/65.3 | 114 | 34.5/27.1 |
| 13 | 12 | 278 | 78.8/71.4 | 350 | 87.5/79.1 | 137 | 40.7/33.3 |
| 15 | 14 | 325 | 91.5/84.1 | 409 | 101.6/93.2 | 160 | 47.0/39.0 |

*Source*: PIANC
*Note*: Coefficient (n) is the ratio of channel wetted surface cross-section to barge immersed cross-section.

minimum curve radius of 800 m, 10 times the standard barge length.

Additionally, navigation can be restricted by artificial structures such as bridges or power cables which reduce overhead clearance ('air draught'), underwater cables or pipelines reducing water depth and marginal structures such as bridge piers or wharves which reduce width of the available channel. Lock gates will always be a critical determinant of overall vessel size and the design standards for new or improved lock gates must be planned with great care in relation to anticipated future demand and optimum vessel size. All too easily lock gates can become a restriction on navigation which will be costly to remedy.

## IMPROVED RIVER NAVIGATION

It will be apparent from the above discussion that there is considerable scope for improving one or other or indeed several of the physical characteristics of rivers. It is well for planners to keep in mind the complex relationship between the different parameters and the possible knock-on effects of modifying any one of them. For example, velocity, channel form and sediment transport are closely related. Few of the rivers used for navigation have not been improved in some way (Hadfield, 1986) and a river such as the Rhine is now more artificial than it is natural.

The most usual engineering works for river improvement relate to water depth and the scale can vary from the local to the regional and the cost from moderate to vast. The Zaire River was improved in the 1960s by blasting local high points in a relatively small number of places, while on the Parana–Tiete river network in Latin America a regional scale improvement involves a succession of large dams which, when completed, will provide a 2,600 km navigation network for four-barge push–tows of 6,000 tonnes capacity. This is seen as a way of facilitating exports of grain, sugar cane, cattle, soya beans and distilled alcohol from large areas of São Paulo, Minas Gerais and Goias in Brazil and from Paraguay.

On a smaller scale, but still of considerable significance locally, has been Asian Development Bank funding for Indonesian projects to improve river navigation in Kalimantan (inland from Bandjarmasim and Samarinda) to allow the use of larger craft for longer periods of the year and thereby improve the cost effectiveness of river transport in an area where other transport is poorly developed. A possible river for improvement could be the Fly River in Papua New Guinea, on which five 3,000 tonnes capacity self-propelled barges are used to transport copper ore from the Ok Tedi mine to the coast for export but where the flow regime is very unpredictable, not to mention opposition from local inhabitants which takes the form of bow-and-arrow attacks on barges!

Water depths can be increased either by dredging, often a short term and temporary solution, or by the construction of barrages or weirs which raise the water level behind them but which need built-in navigation locks for the passage of vessels. Clearly this is a more permanent but also more costly solution but also provides dimensions for craft that cannot easily be altered at a later date. Built in the mid-ninteenth century primarily for irrigation purposes the Damietta and Rosetta barrages on the Nile were provided with locks and greatly improved navigation. The locks were upgraded in the 1890s. The Asyut, Nag Hammadi and Isna barrages also served to increase water levels and were provided with small locks (70×15 m) and the original Aswan dam of 1925 had locks of 80×9.5 m. However, the new Aswan dam of 1971 has no such locks and therefore interrupts navigation which has to be resumed above the dam on Lake Nasser.

In the 1930s, blowers were used to remove shoals on the Ahrewady, although such improvements were usually very short-lived, while in the 1980s there has been considerable dredging of the Chao Phraya in Thailand. Dutch financial aid and expertise has been instrumental in numerous dredging projects including several on the Ganges and 400 km inland from Calcutta the construction in the 1980s of the Farakka barrage. With its navigation locks and associated canals this allows year-round navigation where previously it was seasonal and the 1,500 km waterway route from Calcutta to Patna and Allahabad has in consequence been designated a National Waterway and the Farakka route avoids a long detour through Bangladesh waterways. Political problems have unfortunately delayed use of the new route. The Kottapuram–Quilon canal in Kerala is to be improved with Indian government funding and private investment is being sought by the Kerala government for improvement of the Udyogamandal canal and a number of wharves.

In China, the Changjiang (Yangtze) has seen massive improvement works since the 1950s. From Chongqing to Lingxiang the depth has been increased from 1.8 to 2.9 m and from Lingxiang to Hankou from 1.7 to 3.2 m and the navigable channel width increased to 60 m and curve radii to 750 m (Min. of Comm., 1965; Qiyu, 1987). This allows the use of 10,000 tonnes capacity barge tows from Hankan to Lingxiang, 6,000 to 8,000 tonnes between Lingxiang and Yichang and 3,000 tonnes up to Chongqing. Hundreds of dangerous shoals have been removed by blasting and dredging and in 1981 a new HEP dam was completed at Gezhouba, one of a number of multi-purpose schemes, which raised water levels for some 200 km upstream in a section of the river in which previously navigation had been particularly difficult. Tributaries and lakes such as Dongting and Pyang have also been dredged but there are still many places where shore winches have to be used to pull vessels through rapids. There are plans to reduce channel gradient to eliminate the need for such winches and also to improve the main channel throughout to 3.2 m depth, 70 m width and 850 m curve radii. There are still some 41 sections of the river in which navigation is restricted and where local improvements are gradually being undertaken.

On a completely different scale are proposals for the Three Gorges area above Gezhouba which have been under consideration for 50 years or more (Zuogao, 1990) and which would involve impounding the river to about 100 m above its present level and creating a back lake to Jiangjin and Chongqing. The vast dam to effect this would have to be provided with twin, multi-step locks, presumably of dimensions comparable with those at Gezhouba (120×18×3.5 m) and construction time has been estimated at 18 years with some 21 years for the river to reach its new level of 175 m above sea level. Given the vast size of its waterway network, China seems determined to maximise the transport potential although there have been unfortunate cases where barrages built for flood control, irrigation or HEP have not been provided with the necessary navigation locks and the total length of navigable waterway has in fact been reduced from a peak of 172,000 to 109,000 km (Qiyu, 1987).

On the Senegal river in West Africa, the Diama barrage on the lower part of the river, built essentially for saline exclusion and irrigation, has been equipped with navigation locks (190×25 m) and will increase the assured river depth to 1.5 m as far inland as Kayes on the Mali border. This has great potential for land-locked Mali for which both normal and exceptional drought relief traffic has been seriously hindered by the inadequacy of the rail link from Dakar to Bamako. Dredging to 1.8 m is planned for the Lokoja–Onitsha section of the lower Niger to facilitate transport to and from an iron and steel production complex which is under construction at Ajaokuta.

The Parana–Tiete improvement has already been mentioned. The Parana is navigable for sea-going ships to Rosario with smaller craft (Figure 2.2) able to navigate to Asuncion on the Paraguay and to Itaipu on the Parana where a large HEP dam, for which navigation locks have been planned but not yet built, provides the head of navigation. The Tiete tributary of the Parana has its source close to the Atlantic coast in Sao Paulo state and seven dams for HEP are being provided with locks (136×12 m) and barges of 2,000 tonnes will be able to navigate from Conchas to the Parana confluence and, were the Itaipu locks to be constructed, all the way to the sea.

In Minas Gerais and Bahia states the Sao Francisco river is navigable between Pirapora and Juazeiro and just above the latter is the Sobradinho dam and lock (Greer, 1984). This lock can take craft of 120×17 m and with a lift of 33.5 m is one of the world's deepest. The 340 km of the lower Orinoco has been dredged to the confluence of the Caroni which has become a main transhipment point for iron ore from barges to sea-going vessels of up to 100,000 tonnes.

Channel straightening is not infrequent but usually localised and not on a large scale. A possible exception is the Jonglai canal in southern Sudan which, although not intended primarily for navigation, will certainly have the effect of shortening distances and improving navigation. Frustrated by a succession of grandiose studies but all too little action on the dredging of the Paraguay–Parana rivers, a Paraguayan shipowner proposed a different approach. 'The fundamental thing is to adapt the boats to the river, not the river to the boats' (*Fairplay*, 27/12/1990).

## NAVIGATION AIDS

Because of the physical characteristics of many rivers, navigation in daylight is often difficult and at night impossible. This clearly reduces the capacity of waterways for transport and increases the time and cost. However, aids can be provided which reduce the risks and improve navigability. A distinction is sometimes made between aids to navigation which are located on the shore or in the channel and those which are in the form of equipment carried on the ship itself. The basic aim of all such aids will be to keep the vessel in the fairway or channel and avoid collision with other craft. The channel itself can be marked with stakes, buoys, lights and radar reflectors and 'leading marks' can be placed on shore which enable vessels to adopt appropriate approach lines. Radar has greatly improved navigation and is now available at a wide range of prices but is still far from universal in use especially on smaller craft.

While radar may provide a suitable picture of land margins to water courses and of the buoys marking the channels it is often the case that the water–land margin is far from distinct in low-lying marsh areas or deltas and the channels do not always have suitable markers. There is considerable scope for providing navigation aids but their value depends on the accuracy of initial surveys and continual monitoring of channel configuration – not an easy or cheap task for the waterways of Bangladesh, the Ahrewady or, indeed, many longer tropical rivers where channel changes are frequent, rapid and dramatic in character. There is the possibility of phased improvement of navigation aids related to growing demand for inland navigation or the need to use larger vessels and a number of countries have classified their waterways with respect to navigability and aids.

In Bangladesh the Class I waterways are surveyed, have guaranteed water depths of 1.8 to 3.6 m, have channel marks, buoys and beacons and are navigable at night. The Class II waterways have depths of 1.4 to 2.7 m, will have been partially surveyed and may have some navigation aids but no guaranteed standard. The Class III water courses are neither surveyed nor marked. On the Chinese waterways, Class I denotes a continuous, buoyed, lit channel, Class II will be marked in heavily used sections and Class III have scattered marks and are not lit. Inland navigation, wherever it is practised, is highly dependent on local expertise and knowledge is usually passed on in informal ways and rarely exists in documentary forms. The skill involved in navigating many of the rivers is not to be underestimated.

## INLAND WATERWAY CRAFT

Earlier in this chapter it was suggested that waterway transport is characterised by an almost infinite variety of craft in terms of size, design, technical sophistication, construction materials, propulsion and carrying capacity. In a very real sense the boats are adapted to the rivers and reflect local geographical conditions and the term 'country boat' is appropriately used for a wide range of traditional craft. In many Developing Countries, Bangladesh provides an obvious example, the construction of such vessels is still a thriving industry (Jansen *et al.*, 1989). By the very nature and dispersion of these vessels statistical information about their numbers, operations and productivity is always likely to be vague but this in no way detracts from their overwhelming importance in local transport in areas such as Amazonia, the Niger delta and Bangladesh.

Perhaps because of their obvious significance in a country such as Bangladesh, with its dense, intricate network of waterways, the country boats have attracted far more attention from academics and transport planners (Bari, 1982; ESCAP, 1984; Greenhill, 1971; Jansen et al, 1989). In Bangladesh there are over 720,000 country boats with a total carrying capacity of over one million tonnes but only 68,000 with a carrying capacity of 836,000 tonnes are officially classed as commercial operations. The average vessel capacity is 1.5 tonnes, most are of wooden construction, sometimes partly decked and mainly propelled by sails and/or oars. There are about six principal types but as many as 160 regional names for different styles (Bari, 1982). These craft are not subject to government regulation, they are highly competitive, can carry virtually any available cargo literally to and from the farm 'gate' and provide both a local and inter-regional service.

Most of these craft capitalise on local material and expertise to reduce to almost zero the import input. Their fuel requirements are negligible. However, they do offer scope for improvement and in places the size and capacity could be increased, more efficient sail designs could be adopted and many could be modified for motor-isation – indeed, this is often the first step in the improvement process. Outboard motors or the 'long-tail' style (often a road-vehicle motor with direct shaft drive over the stern of the boat) favoured in Thailand are more flexible and one-third of the cost of a full inboard engine. Vessels can be towed in strings and some designs might even be suitable for pushing. While the idea may seem strange some of the designs could be constructed in more durable, and possibly cheaper, ferro-concrete. There have been factories making such boats at Juba in Sudan and also in China. Most improvements will increase the import element both at the construction stage and also in operation (e.g. fuel, spares) and this should always be considered with respect to the sustainability of the technology – the traditional technology had the considerable advantage of being almost infinitely sustainable.

Motorised commercial craft in Bangladesh total well over 2,000 units with a combined cargo capacity of over 350,000 tonnes and include over 800 passenger vessels with a total capacity of 94,000 persons (Bari, 1982). Most of these craft are best described as launches but the numbers certainly include many small motorised craft operating at the local level. The fleet also includes around 500 'dumb' barges which have to be towed.

Given its environment, it is not perhaps surprising that in Bangladesh an estimated 65 per cent of the total freight movement and 38 per cent of the passenger traffic is by water. It is a country heavily dependent on waterways yet the shallowness of much of the network and the highly dispersed and low level of local demand mean that larger vessels and more sophisticated technology are not everywhere justified although undoubtedly there is considerable scope for selective upgrading. India too has

*Plate 2.2* Waterways, as in Bangladesh, provide natural arteries for movement, facilitate bulk transport and the style and variety of craft will reflect local geographical conditions (Photo: Intermediate Technology).

a vast number and great variety of country boats (10,000 registered with state authorities, so vastly more certainly exist) and on National Waterway Number 1 from Calcutta inland more sophisticated push–tow craft have been introduced and there have been experiments with small tow units on the middle Ganges. Tugs and barges are also found in Assam, Maharashtra , Kerala and Goa.

In China over 93 per cent of all inland waterways craft are non-motorised (Qiyu, 1987) but it has very large numbers of engined craft and with the size of its main waterways and the great demand for long distance movement of bulk commodities (coal, ores, grain) has introduced high capacity push–tow technology on some of the main routes. In 1981 the Dravo Corporation of America provided four 6,000 hp tow-boats and 30 barges of 1,500 and 2,000 tonnes capacity and integrated barges for push–towing of 5,000 tonnes have also been introduced. China is able to capitalise on economies of scale to a degree few Developing Countries are able to match and sophisticated inland shipping technology is now common on the Changjiang, Huaihe, Songhuajiang and some other rivers. The Changjiang River Shipping Company alone now controls over 200 tugs/pushers and 1,500 barges.

On rivers such as the Mekong the technology is generally less sophisticated but still very important for local and longer distance movement. Shawcross (1984) describes Bentre on the Mekong as 'a dockside crowded with boats – long wooden passenger boats . . . small sampans on which peasants had brought their vegetables, ducks and pigs . . . floating homes . . . ancient ferries'. Thailand has an estimated 7,000 wooden craft only 1,800 of which exceed 60 tonnes capacity and 750 steel barges of 100 to 300 tonnes restricted to the Chao Phrya. On Ghana's Lake Volta a low-draught, self-propelled vessel with bow ramp for fork-lift truck loading of bagged agricultural products provided effective transport over the 400 km length of the lake and in 1993 Danish Government aid was provided for vessel and pier rehabilitation.

## PUSHING VERSUS PULLING

Over time, rowing boats have been used to tow large craft, especially sailing vessels left becalmed, and the advent of steam propulsion clearly made possible the regular towing of much larger individual units or numbers of them. Use of a tug and barges had the clear advantage that the propulsion unit, the tug, and its crew could operate the barges on a drop-and-swap basis and thereby maximise the utilisation of the more costly part of the system by separating it from the cargo carrying units. The propulsion unit was not detained while the cargo was handled into and out of the barges and this became the basis of many lighterage operations (e.g. on the river Thames).

A barge, and particularly a number of barges towed behind a tug in a long string on tow ropes, does not constitute an easily manoeuvred unit – it has difficulty negotiating bends, is not easily stopped and the energy efficiency of the system is low. A more rigid system clearly has advantages and a common way of achieving this was to lash the tug alongside the barge, or possibly have a barge on either side. The barges do not necessarily have to be of the same dimensions. The next step was to place the tug behind the barge so that it became a pusher – this reduces the resistance, improves the energy efficiency and manoeuvrability and also led to the idea of modular barges which could be coupled together in various configurations (Figure 2.3).

Push–towing was experimented with in Europe in the nineteenth century (Hadfield, 1986) but it was in America on the Mississippi that the technology was developed most fully and on a scale not found elsewhere. By the 1970s tow sizes of 50,000 tonnes were not uncommon and one powerful tow-boat (the American pusher-tug) has been known to handle as many as 70 empty barges as one unit. In Europe in the 1960s push–tow technology spread rapidly and has become common on larger waterways with two, four and six barge configurations now the norm (Figure 2.3). The actual way in which the barges are grouped will be determined by the ruling dimensions on particular waterways – bridge widths, lock gate sizes, curves and safety considerations. It was only after lengthy experimentation that the Dutch government permitted six-barge tows which had been operating on the German part of the Rhine for a number of years.

The Belgians introduced push–towing on the Zaire (Congo) River, where such units still operate and push–tow technology, with barge module sizes related to local conditions, is found on the Niger, Gambia, Ganges and a number of Chinese rivers. In 1995, the Compania Paraguaya de Transporte Fluvial of Asuncion took delivery of a 43 m, 4,500 hp tow boat to operate with new 61 × 16 m barges to transport grain on the Paraguay River. The tow-

| Classes of navigable waterways | Motor vessels and barges — Type de bateau: Caractéristiques générales / Type of vessel: General characteristics | | | | | Pushed units — Type de convoi-Caractéristiques générales / Type of pusher- General characteristics | | | | | Push lighters — Type de barge: Caractéristiques générales / Type of barge: General characteristics | | | | Minimum height under bridges |
|---|---|---|---|---|---|---|---|---|---|---|---|---|---|---|---|
| | Dénomination Designation | Longueur Length | Largeur Beam | Tir. d'eau Draft | Tonnage Tonnage | Dénomination Designation | Longueur Length | Largeur Beam | Tir. d'eau Draft | Tonnage Tonnage | Dénomination Designation | Longueur Length | Largeur Beam | Tir. d'eau Draft | |
| (1) | | III | III | III | T | | III | III | III | T | | III | III | III | III (4) |
| I | Péniche Barge | 38.50 | 5.05 | 2.20 | 250-400 | | | | | | | | | | 4.00 |
| II | Kast Campinois Campine-Barge | 50-55 | 6.50 | 2.50 | 400-650 | | | | | | | | | | 4.50 |
| III | D.E.K. | 67-80 | 8.20 | 2.50 | 650-1000 | | | | | | | | | | 5.00 |
| IV | R.H.K. | 80-85 | 9.50 | 2.50 | 1000-1500 | 1 barge E I | 85 | 9.50 | 2.50 | 1240 | Europe I | 70.00 | 9.50 | 2.50 | 5.25 |
| V a | Grands Rhénans Large Rhine Vessels | 95-110 | 11.40 (2) | 2.80 | 1500-3000 | 1 barge E II | 95-105 | 11.40 | 2.80 | 1850 | Europe II | 76.50 | 11.40 (2) | 2.80 | 7.00 |
| V b | | | | | | 2 barges E II | 172-185 | 11.40 | 2.80 | 3700 | | | | | |
| VI a | | | | | | 4 barges E IIa | 185-195 | 22.80 | 4.50 (3) | 8000 12000 | Europe II a | 76.50 | 11.40 (2) | 3.90 | 9.10 |
| VI b | | | | | | 6 barges E IIa | 270 / 195 | 22.80 / 34.20 | 4.50 (3) / 4.50 (3) | 12000 18000 / 12000 18000 | | | | | |

*Figure 2.3* European push–tow configurations
*Source:* PIANC (1990)

*Note:* 1) The class of a waterway is determined by the horizontal dimensions of the vessels or pushed units.
2) In the Danubian basin, this beam is usually 11m.
3) Takes into account the future developments.
4) Takes into account a security clearance between the draught of a vessel and the height under the bridges. In case of container traffic, the height under the bridges should be checked.

*Table 2.2*  The barge-carrier systems

| | Length (m) | Beam (m) | Loaded draught (m) | Number of barges | Number of containers | Deadweight tonnage |
|---|---|---|---|---|---|---|
| | | | *a) Typical dimensions of 'mother' ships* | | | |
| LASH (i) | 261.4 | 32.5 | 12.1 | 83 | — | 48,303 |
| LASH (ii) | 261.4 | 30.5 | 10.7 | 60 | 510 | 30,298 |
| LASH (iii) | 272.3 | 30.5 | 11.6 | 89 | 510 | 41,578 |
| SEABEE | 267.0 | 32.3 | 11.9 | 38 | — | 39,026 |
| RUSSIAN LASH | 266.5 | 35.0 | 11.0 | 26 | — | 38,000 |
| BACO-LINER | 204.1 | 28.5 | 6.7 | 12 Baco or 14 LASH | 500 | 21,000 |
| BACAT[a] | 104.0 | 20.7 | 5.4 | 3 LASH plus 10 Bacat | — | 2,500 |
| SPLASH[b] | 149.9 | 33.9 | 4.9 | 19 LASH | — | 10,180 |

| | Length (m) | Beam (m) | Loaded draught (m) | Cargo capacity (tonnes) |
|---|---|---|---|---|
| | | *b) The barge units* | | |
| LASH | 18.75 | 9.5 | 2.73 | 370 |
| SEABEE | 29.7 | 10.7 | 3.2 | 840 |
| BACO | 24.0 | 9.5 | 4.1 | 800 |
| BACAT | 16.8 | 4.6 | 2.6 | 147 |
| DANUBE/ RUSSIAN | 38.2 | 11.0 | 3.3 | 1,100 |

*Notes:* (a) BACAT – barge catamaran system designed to link British with mainland European waterways

(b) Now operated in Chinese waters

boat has been specially designed to operate under conditions in which there are only a small number of refuelling points available and large bunkers are necessary.

## BARGE-CARRIER SYSTEMS

In 1969 on the North Atlantic a vessel was introduced with the ability to lift barges aboard at New Orleans for an ocean journey to Rotterdam where they could be put back into the water for their onward journey. This was the birth of the Lighter Aboard Ship (LASH) technology and during the 1970s a number of design variants (Table 2.2) were introduced (Hilling, 1977a). The obvious role, and the function which inspired this innovation in technology, was that of linking inland waterway systems across intervening maritime space and there seemed particular advantages where such waterways were in Developing Countries in which the deep-sea port facilities are often inadequate (Chapter 7). The Barge Carrier Vessel ( BCV) merely had to anchor mid-stream to handle its barges into or out of the water and the actual cargo handling could be done at shallow water facilities at ports or further inland on rivers or canals. The areas identified as suitable for BCV operations (Table 2.3) include many of the river systems already mentioned in this chapter – Amazon, Parana-Paraguay, West African rivers, South-East Asian rivers (along which are vast concentrations of population) but also archipelago and 'enclosed' sea areas.

*Table 2.3* Geographical conditions suitable for barge-carrier systems

| | |
|---|---|
| (1) Inland waterways | |
| a. United States | – Mississippi, Arkansas, Ohio, Snake, Intra-coastal waterway |
| b. Europe | – waterways Class IV and better – rivers – Rhine, Danube, Elbe, Rhone, Seine, Thames |
| c. Amazon basin | |
| d. Parana/Paraguay | – improvement in progress |
| e. West Africa | – Senegal, Niger, Gambia, Ogooué, lower Zaire |
| f. South East Asia | – Hooghly, Ganges, Salween, Ahrewady, Mekong |

Such waterways, as in South-East Asia, serve densely populated areas, are in many cases already used intensively for transport and either are or could be further improved to take larger craft.

| | |
|---|---|
| (2) Coastal/oceanic feeders | |
| a. Archipelago areas | – Indonesia, Philippines, South-West Pacific generally, West Indies, Aegean |
| b. Enclosed sea areas | – Arabian Gulf, Red Sea, Gulf of Mexico, Bay of Bengal, Baltic, Mediterranean |
| (3) Estuary, sheltered waters | – San Francisco Bay, Chesapeake Bay, Puget Sound, Thames estuary, Tokyo Bay, Osaka Bay |
| (4) Intra-port lighterage | – Singapore, Hong Kong, London, Hamburg, Rotterdam, all 'primitive' ports with no deep-water berthage |

*Source:* Hilling (1977)

Other advantages of the BCV system for Developing Countries were that it could act as a bypass for congested ports, it could provide a more flexible cargo-handling system and it allowed the use of labour-intensive cargo handling methods without adding to the time, and therefore the cost, of keeping the ocean-going ship in port. It could also be argued that the inland transport routes were often unsuitable for containers and many of the bulk or semi-bulk primary product exports were better suited to movement in barges than they were in containers – logs and sawn timber provide obvious examples. The modular barges were highly suitable in terms of capacities and dimensions for a wide variety of bulk goods in smaller quantities (some ores, concentrates, fertiliser), forest products (logs, sawn timber, veneers) and bagged agricultural produce – often exported in less than ship load quantities but more than container load. Much of such trade is seasonal and expensive port facilities such as are needed for containers may not therefore be justified. Barges can also provide valuable additional storage which is often needed but not always available at the coastal and inland transit centres.

A study of Indian trade (Indo-American Chamber of Commerce, 1975) concluded that containerisation required a wholesale restructuring of the transport infrastructure at a cost that could not be justified whereas LASH had a number of advantages:

1 It is a more flexible method of introducing unitisation (see Chapter 7) alongside existing conventional cargo handling and would allow two-way, import and export, employment of capacity.
2 The LASH system allows the fullest possible use of the existing cargo-handling facilities, reduced capital investment at ports and is particularly useful where berth draught has been reduced by siltation (e.g. Calcutta – but this is a problem in many tropical ports). LASH also makes it possible to utilise normally shallow-water, small-ship berths and distribute traffic to small ports on varied waterways and possibly well inland. This helps to counteract an undesirable tendency towards concentration in the modern sector.
3 Most ports in India, Bangladesh, Sri Lanka, Pakistan and Indonesia had still to be modernised and LASH reduces the need to divert scarce resources away from directly productive activities which form the basis of economic development (see discussion in Chapter 1).

4 Heavy seasonal monsoon rains disrupt cargo handling and cause considerable delay for conventional vessels. By diverting the cargo handling from the ship to varied locations the likelihood of delay for the ocean carrier is reduced.

In these respects, what was true for India was also true for many Developing Countries and at the peak of its development in the late 1970s the LASH and other BCV systems were serving a large number of ports in Developing Countries in Latin America, Africa and Asia. Yet it has to be admitted that the BCV system did not take off as its supporters, including the author, had anticipated.

There were a number of reasons why there was a gap between the perceived advantages and the actual adoption of BCV technology. A major factor was undoubtedly the adoption and spread of containerisation at a much faster rate than expected with many Developing Countries obliged to adopt the system and improve their ports accordingly and irrespective of the financial and social costs. In the 1980s there was general recession in world trade and the pressure on ports was reduced. The first generation BCVs were large and very expensive and shipping companies were reluctant to

*Plate 2.3* On the Parana river in Paraguay, a low-draught, high capacity push–tug and barges was introduced in 1995 by the international Cargill company to move grain for export (Photo: Cargill plc).

take the risk. On many routes, even those where onward movement of barges could be by way of inland waterways, it was often
the case that the barges did not move beyond the ports – thereby
removing the principal raison d'etre of the whole system.

For a number of years in the late 1970s and early 1980s the
smaller BACAT technology, designed to link the waterways of
Britain and mainland Europe, provided a valuable link between
congested ports in the Arabian Gulf and Bombay, with the sea-
going journey of the mother ship providing the additional time
needed for the loading and unloading of the barges at either end.

Significantly, BCVs still operate very successfully as a link
between North-West Europe and West Africa, a service operated
by three BCVs of the BACO (BArge-COntainer) design belonging
to the Rhein Maas See company, and also from America and
Europe to the Middle East and Asia. BCVs were withdrawn
from South American trades but there was talk in 1994 of their
reintroduction because of problems at some ports. In 1993, there
was an increase in timber and drummed latex out of Port Kelang
by barge. The Russians have taken over from the Americans as the
principal operators of BCV technology with services from Black
Sea ports to India and South East Asia. The Russians have also
been trying to extend BCV operations in Thailand where the Chao
Phrya river provides an obvious artery for onward movement of
barges. The Chinese have acquired several smaller 'feeder' BCVs of
the SPLASH variety (Table 2.3) originally belonging to the American Central Gulf Lines and are using them on their rivers and also
to link up river systems by a coastal service. The BCV system
worked effectively in Bangladesh and barges taken off the mother
ships at Chittagong were then towed to Khulna or Dhaka. There
are numerous other river systems which could be linked to good
effect by BCVs and it would certainly be premature to write off
this form of inland waterway transport.

## RIVER PORTS

For cargo handling and the embarking and disembarking of passengers some form of wharf is required – for small craft this may
be a very simple structure but where large volumes of bulk
commodities are being handled or large numbers of passengers
dealt with the terminals may have to be very elaborate.

In Bangladesh there are over 1,300 recognised handling places ('ghats'), mainly for passengers and often no more than a simple wooden staging, either marginal to the river bank, a wharf or quay, or projecting from the bank into the river, a pier. Such structures are perfectly adequate where little cargo is handled and where the water depths do not vary greatly by reason of tidal changes or seasonal flow patterns.

However, on some Chinese rivers, water levels have been known to change by as much as 10 m in 24 hours and it is therefore necessary to have a wharf design which can accommodate such changes. Traditionally, Chinese wharves were of the 'marginal' type, a staging built along the river bank, but especially in places where cargo volumes increased, where uniform flow of cargo was important or where more sophisticated cargo handling methods were thought necessary, it became desirable to develop wharf designs that allowed unhindered use at all levels of water. Five main types may be identified (Figure 2.4).

In the lower courses of rivers where the variations in water level are less pronounced it is usual to have the 'vertical' type of quay which is flexible for bulk cargo handling and for use as a multi-purpose terminal (ESCAP, 1982). These are found widely in Harbin, Jiamusi, Jiangsu and Zhejiang provinces. With wider variations in river level, for example the middle Changjiang, the 'inclined' style is more common. For passengers these create no problems and smaller craft can ground for short periods on the slope but cargo handling can be a problem if the slopes are too steep. The position of the floating pontoons have in cases to be adjusted as water levels change and this can interrupt cargo handling. High-velocity river flow in flood conditions and ice on northern rivers can put pressure on the pontoons and can create operating problems. In some designs the whole wharf and the link span connecting it to the shore is on a track system which allows movement up and down slope. The 'floating' type automatically adjusts to changing water level and is therefore more efficient than the inclined type. Where the pontoon has to be at some distance from the shore to remain at the channel edge it may be necessary to have several link spans to the shore. There can be problems of weight support where the link spans have to be long.

A variety of methods are used for moving the cargo up and down the slope or link span including rail tracks with small wagons, conveyor belts and even cable cars. For ease of cargo handling the vertical wharves are clearly preferable allowing as they do a direct

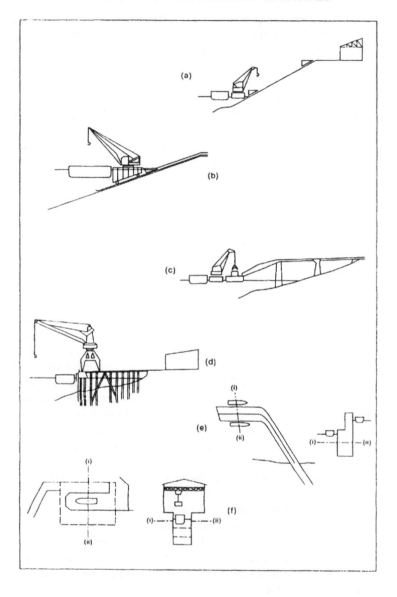

*Figure 2.4* Chinese river quay designs
*Source:* Modified from ESCAP (1982)
*Notes:* (a) and (b) 'inclined' type   (c) 'floating' type   (d) 'vertical' type
(e) 'double-step' type   (f) 'dredged-basin' type

lift on/lift off movement and where these have replaced the inclined types the cargo handling productivity has increased by as much as 15 per cent (ESCAP, 1982). Where the water level changes are greater, as in the upper reaches of some rivers, and where the flood period is short, the double-step vertical wharf may be used. Some semi-vertical types have also been found cheaper than inclined forms.

In areas with long periods of heavy rain which can seriously disrupt cargo handling and result in delay and congestion (e.g. Changsa in Hunan) a dock basin may be built into the river bank and provided with a sheltering roof. This solution has still not been widely adopted and is certainly not possible where water level changes are excessive.

China has over 1,000 river ports, 45 per cent of which handle less than 500,000 tonnes a year, 44 per cent are in the 500,000 to two million tonnes range, eight per cent handle from two to five million tonnes and three per cent handle more than five million tonnes (ESCAP, 1982). Given the vast size of its waterway system and the

*Plate 2.4* A simple ramp allows ease of loading palletised rice by fork-lift truck at Ghana's Volta Lake port at Yapei, but on waterways of variable depth more sophisticated solutions may be necessary.

number of substantial river ports it is hardly surprising that China, more than any other Developing Country, has given great attention to the design and engineering aspects of providing efficient cargo handling terminals. Mechanical handling now accounts for 50 per cent of the cargo. Since 1949, the number of terminals on the Changjiang has doubled and countrywide has tripled. China provides the best example of a country in which the real potential of inland waterways transport has been recognised and while much remains to be done the progress has already been impressive.

In China and elsewhere the level of sophistication of the riverside landing places will determine the size and type of craft that can be accommodated, the types of cargo that can be efficiently handled and the cost of so doing. Not surprisingly, port improvement has in recent years been a feature on many waterways including the Amazon, Paraguay (Hayman-Joyce, 1979), Niger and Zaire.

## PLANNING IMPLICATIONS

In mature and Developing Countries alike there are sound reasons for suggesting that water transport should be used to the maximum possible and the case is strengthened as environmental considerations become ever more critical. On broad environmental grounds, water transport produces little noise, relatively low levels of atmospheric pollution and vibration and serves to reduce heavy vehicular traffic on road systems that in many places are already congested or costly to maintain at satisfactory levels. Water transport is especially valuable in the movement of outsize loads and also hazardous commodities. While the environmental considerations may not seem to have the same immediacy in many Developing Countries that they do in the mature economies there is no reason why they should be considered of less importance – there are compelling arguments for not following the route charted by the mature economies and many Developing Countries themselves already exhibit aspects of transport induced environmental degradation which could be checked before they assume unmanageable proportions.

For Developing Countries there may be more important factors in favour of water transport. The tonne-km cost of water transport will vary greatly depending on the circumstances but will often be less than for other modes in part because of the low import element in the total cost. Of considerable importance is the ability of water transport to benefit from what has been called the 'cube

*Table 2.4*    Inland shipping cost break down by vessel size (ECUs, 1987)

|  | Self-propelled vessels (dwt) | | | | | Push–tow |
|---|---|---|---|---|---|---|
|  | 300 | 600 | 1,200 | 1,600 | 2,500 | 10,000 |
| Total cost | 97 | 131 | 214 | 260 | 359 | 2,362 |
| Average cost per tonne | 0.27 | 0.21 | 0.17 | 0.16 | 0.14 | 0.23 |
| As a percentage of total annual cost: | | | | | | |
| Capital | 9.3 | 16.0 | 28.5 | 28.1 | 23.7 | 18.0 |
| Labour | 67.0 | 55.7 | 38.8 | 36.9 | 35.4 | 27.6 |
| Fuel | 12.4 | 15.3 | 16.8 | 18.5 | 20.6 | 32.3 |
| Other | 11.3 | 13.0 | 15.9 | 16.5 | 20.3 | 22.1 |

*Source*: After PIANC (1991)

law' whereby a doubling of the vessel's dimensions leads to a cubing of the carrying capacity. This allows water transport to benefit from economies of scale not available on roads or railways where unit dimensions are rigidly defined. In water transport the crew and vessel costs do not increase in proportion to size (Table 2.4). Water transport therefore becomes increasingly cost-effective with larger volumes and over greater distances. While the nature of the individual cost curves may vary and in practice may well be stepped, Figure 2.5 indicates the general relationship between cost

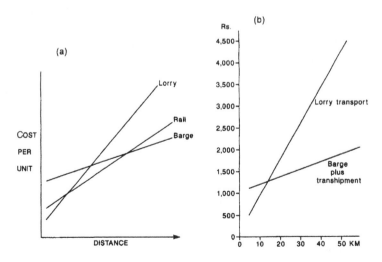

*Figure 2.5*    Modal cost comparisons with distance
*Source*:    After Hoover (1948), Saggar (1979)

and distance for the different modes. Water transport comes into its own over the longest distances when the economies of scale and low line-haul costs outweigh the higher initial capital costs.

The actual cost of water transport will vary greatly from place to place being a function of the type of vessel that can be used (size, method of propulsion, speed), the type and amount of goods transported, the type of enterprise (number of ships, way they are utilised), need for and efficiency of transhipment and the character of the navigation (locks, lock service time, air draught, current velocity, operating hours). The operational costs of a vessel may be identified as its capital, crew and fuel costs. Additionally there will be costs associated with loaded/empty time and possibly cargo handling (PIANC, 1991).

The ideal conditions for inland waterway transport are those in which the goods can be moved from origin to destination entirely by water and without delivery or collection by some other mode (e.g. movement of coal direct from a mine to an electricity generating station). Over short distances the cost of transhipment can eliminate the advantages of water transport. Over increasing distances higher proportions of transhipment cost (including the cost of any other mode needed to start or finish a journey) can be absorbed by water transport without losing its competitive advantage. Table 2.5 gives a cost comparison based on European costs for a 500-tonne door-to-door road haul and a 600-km inland shipping movement with an additional 20-km road haul to the destination. The transhipment costs are particularly significant in the case of general cargo but there are clearly strong arguments for maximising the efficiency and minimising the cost wherever possible.

Because of increasing road congestion in the Bombay region and the availability of a network of rivers and creeks a study was undertaken to assess the relative costs of road and barge transport (Saggar, 1979) based on the use of a tug and four 125-tonne barges. Allowance was made for the cost of transhipping cargo where necessary and while the rates do not show the tapering with distance that might be expected, the low line-haul costs of the barge system give it an advantage with distance greater than 15 km. Over a 20-year period the capital investment in the barge system was shown to be far lower than for road transport – an obvious factor being the higher life expectancy of a barge compared with a road vehicle. Arguably, in many Developing Countries the condi-

*Table 2.5*  Transport cost comparisons

### (a) *European cost comparisons (ECUs per tonne)*

|  |  | Road haulage | Inland shipping with additional road haul |
|---|---|---|---|
| Fodder (bulk) | Road | 19.0 | 2.5 |
|  | Ship | — | 4.5 |
|  | Transhipment | — | 3.0 |
|  | Total tonne cost | 19.0 | 10.0 |
| General cargo | Road | 18.0 | 4.5 |
|  | Ship | — | 7.5 |
|  | Transhipment | — | 10.0 |
|  | Total tonne cost | 18.0 | 22.0 |

### (b) *Indian cost comparisons (Paise per tonne-km)*

| Distance (km) | Rail[a] Coal | Rail[a] Fertiliser | Road Coal | Road Fertiliser | Barge[b] 500[c] | Barge[b] 1,000[c] | Barge[b] 1,500[c] |
|---|---|---|---|---|---|---|---|
| 50 | 44.2 | 41.4 | 24.2 | 25.8 | 13.7 | 10.3 | 9.4 |
| 100 | 25.7 | 23.7 | 18.2 | 19.4 | 9.1 | 6.5 | 5.8 |
| 300 | 12.6 | 11.4 | 14.7 | 15.7 | 6.1 | 4.1 | 3.4 |

*Source*: (a)  PIANC (1991)
        (b)  Dalvi and Saggar (1979)
*Notes*: (a)  Single line, diesel traction, wagon load
        (b)  Self-propelled, 75 per cent load factor
        (c)  Carrying capacity – tonnes

tions of the roads and the problems associated with proper vehicle maintenance serve to widen this contrast. Similar results have been obtained in Pakistan (Chaudhry, 1987; Khan, 1987).

While the operating cost break down shown in Table 2.4 relates to European conditions certain comments of wider applicability are in order. With self-propelled vessels the economies of scale are significant but these are related to the proportion of capital costs which at first rises but then reduces for the largest vessels and a declining proportion of labour costs. In consequence, fuel and other costs (e.g. cargo handling) as a proportion of the total increase with vessel size. This has implications for Developing Countries in their choice of waterways technology.

For many Developing Countries, fuel availability and cost are critical factors. Indian studies show fuel consumption of 7.8 to 10.0 litres per 1,000 tonne-km for barges contrasted with 40.0 to 46.6 for lorries (Dalvi and Saggar, 1979) Other studies point to

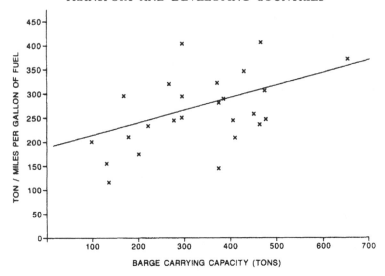

*Figure 2.6* Fuel consumption and barge size
*Source*: After British Waterways Board (1975)

similar conclusions (Baldwin, 1980; Mir, 1987) and a study of fuel consumption by small self-propelled barges (Figure 2.6), craft of just the sort of dimensions that could operate on many smaller waterways in Developing Countries, showed the advantage of using the largest possible craft in any situation (British Waterways Board, 1975). Yet while they may be more economical in terms of tonnages moved, the larger craft clearly require the ready availability of larger quantities of fuel and strategic bunkering facilities. In this respect, the problems of the use of large push–tow units in Paraguay have already been noted.

For non-oil-producing Developing Countries where oil imports often constitute a large element in the import bill and where supplies are all too frequently erratic, there are strong arguments in favour of the adoption of the transport mode which does most to economise on the use of fuel.

Another possible argument for using water transport in Developing Countries is that it can be relatively labour-intensive, especially where country boats are in use. In India, employment generated per 100,000 rupees invested is 33.6 persons per year for water transport, 27.0 for bullock carts, 20.0 for coastal shipping, 10.7 for lorries and only 4.3 for railways (Dalvi and Saggar,

1979). However, the same investment in road maintenance generates employment for 27.0 persons per year. In Bangladesh the country boat sector has always been a source of employment for large numbers but with the adoption of more modern technology for craft and cargo handling the number declines (Jansen et al, 1989). Clearly, much here hinges on the emphasis placed by government on the desirability of job retention or creation.

Water transport has distinct advantages where the tonnages to be moved are large, where the commodities can be handled in bulk and are of a type for which regularity rather than speed of movement is of importance – this applies to coal, oil, ores, fertiliser, some chemicals, aggregates, some building materials, forest products, grains and some non-perishable agricultural commodities. These are often important in both the domestic and external trades of many Developing Countries and the volume/weight relationship calls for the lowest possible transport costs. Any industries based on these materials are best located alongside waterways where they can capitalise on water transport and where transhipment to other modes can be avoided.

While particularly suited to high bulk/low value commodities for which speed is less significant, evidence from North America and Europe (PIANC, 1992; Van den Heuvel, 1993) suggests increasing road congestion and delays and broader environmental considerations are encouraging the transport by barge of higher value goods either in containers or road vehicles. Container movement by barges on the Rhine and Danube has increased dramatically and has also become important on America's Columbia–Snake waterway system. The carriage of road vehicles by barge may seem contradictory but has been of growing significance on the Seine, Rhine and Danube – freight lorries and production cars, lorries and tractors are regularly moved by barge in large numbers. Given the poor state of the roads in many Developing Countries, the inadequacy of the vehicles available and the often woeful inadequacy of the railways (see Chapter 3) there could be an even greater potential for the movement of the whole range of goods by water and this should be given a high priority in overall transport planning.

It clearly makes sense to make the maximum possible use of water transport – it is environmentally friendly, energy conserving and of low cost. The basic infrastructure is natural and the capacity can often be increased greatly at relatively low cost in comparison with other modes. The construction of artificial canals can be very expensive – a recent 3-km stretch of canal in Britain cost £20 million – but it is frequently the case that cost can be shared where there are associated functions concerned with drainage, flood control, irrigation or HEP generation. Ghana's HEP dam at Akosombo on the Volta River created a lake and new 400 km waterway route far into the interior of the country and the cost of turning this into a transport artery was negligible when contrasted with the development of road or rail transport over the same distance (Hilling, 1977b). In 1994, Vietnam, Cambodia, Laos and Thailand announced plans to cooperate in the development of the lower Mekong River for power generation, flood control and transport (Sandberg, 1993) but detailed plans could well depend on the reactions of China and Myanmar who control headwater areas. It is often the case that river basin development, for whatever purpose, is dependent on levels of international agreement that are not always forthcoming.

The great variety and levels of technology available and the associated range of capacities, whether for passengers or freight, allow water transport to be developed in a phased manner in relation to demand – the technology does not have the indivisibility which characterises rail transport either in capacity or investment terms. Perhaps more than other modes over a considerable range of the available technologies there is scope for local input of materials, labour and skills thereby economising on scarce capital and expensive imported equipment and expertise. While water transport tends to be associated mainly with inter-urban transport, at a time of mounting congestion from which Third World cities are certainly not immune, it is well to remember that many of them are riverine and water routes may well have potential for the movement of people and goods within the cities (Hoyle, 1992). Bangkok and Hong Kong provide examples of cities in which good use is already made of water transport and high-speed ferries provide an essential link between Rio de Janeiro and Niteroi – there are numerous other cities in Latin America alone in which ferry services are vital elements in urban transport.

The Pasig River through Manila provides what has been described as an 'under-utilised 10-lane highway' and there has been a recent attempt to develop a river-bus service on which passenger numbers peaked at 750,000 in 1991 but have since dropped back (Flores, 1992). Heavy siltation, the hazard of sunken craft, unsightly river bank settlement, floating garbage and the smell are factors acting against water transport in this case but are problems which can be dealt with as part of an urban improvement programme. A Riverine Transport Development Plan is one element of the Philippine Government's 1993–1998 National Transportation Plan.

If there are places in which the environmental conditions are not ideal it is also the case that water transport has a 'romantic' image largely lacking in other forms of transport. Some railway enthusiasts would doubtless disagree! It was this romantic image which the British Broadcasting Corporation captured in its River Journeys television series and book (BBC, 1984) and which is capitalised upon by the tourist industry in places as far apart as Amsterdam, San Francisco, The Gambia, Thailand, Hong Kong and Egypt. However, there is a danger that water transport is dismissed as being no more than romantic and a possibly valuable contribution to the urban transport system may be ignored. Lagos, Nigeria, provides an example of under-utilised potential while Hong Kong represents a city in which water transport is vital in a highly developed and sophisticated intermodal urban transport system (Hoyle, 1992).

The 'informality' of much water transport may mean that it is largely unregulated and this may well be a factor in the sometimes fatal sinkings of overcrowded river craft. In two separate incidents within a few days in 1990 in northern India over 150 people were drowned and such accidents are all too common. This points to the need for a degree of control and the suggestion that water transport in many places needs to be more fully and more effectively integrated into the overall transport planning and regulatory processes.

# 3

# RAILWAYS – THE INITIATORS OF TAKE-OFF?

It would certainly appear that rail transport has characteristics which attract attention and even generate enthusiasm across a wide range of people – historians, economists, geographers, engineers, politicians and travellers. Perhaps this stems from the vast scale of some rail-road schemes – trans-American continent, Cecil Rhodes' dream of a Cape–Cairo in Africa – or from the sometimes colossal engineering achievement their construction represents.

Some extravagant claims have been made for rail transport. Jefferson (1928) wrote that railways for the century past had 'done more than any single one of man's inventions to transform human life, especially in the way of pushing backward people forward and lifting submerged masses.' For him, the railways had a civilising role and the mobility they provided transformed and ennobled people and had a profound impact on wider aspects of economic development and the evolution of economic and social landscapes.

Jefferson identified a 32-km band, 16 km either side of the track, where the direct influence of the railway was concentrated but Kolars and Malin (1970), in their study of Turkish railways, suggested that the zone of influence may extend up to 40 km either side of the line. This was likely to be variable depending on the particular geographical characteristics of settlement and alternative transport modes.

Meinig (1962) sees railways as a decisive instrument in the creation of many of the significant patterns in human geography and suggests that this holds true to varying degrees wherever railways have been built. He argues that the influence has been least where the railways were superimposed on areas already well

74

developed, more significant where the commercialisation pro-
ceeded simultaneously with development of the rail network and
most dramatic where the railway, as in areas of ninteenth-century
colonisation, was the pioneer mode of mechanised transport.
Overall, few would doubt that railways have had a significant
role in reorganising space (Schivelbusch, 1986). In countries such
as Britain, France and the United States, ninteenth-century industrial and urban expansion went hand in hand with railway construction although a contrast might be drawn between the
European situation, in which the railways facilitated traffic and
the American, in which it created it. In the former the railway
revolution was a consequence of existing industrial production
whereas in the latter it was agriculture and the railways which
provided the starting point.

Meinig (1962) has also suggested that railway networks cannot
be fully understood except in terms of the processes of their
formation – the geographical, economic and political context of
the decision making. In this sense generalisations about railway
systems are not always helpful and Meinig's examination of the
contrasting networks of the Columbia River basin of the United
States and of South Australia illustrates the point. The broad
similarity was the need for agricultural regions to gain access to
seaports for grain exports but in each case the rail strategy was the
victim of changing geographies and there was the influence of local
communities through cash subsidy, land grants and political pressure on route selection. Yet the progress of network evolution was
very different in each case with duplication of routes and active
competition leading to disruption of hinterlands in the Columbia
basin in contrast with lack of competition and stability in South
Australia.

During the main phase of railway expansion in Europe, labour
was cheap but land expensive. In Britain particularly, lines were
built as straight as possible and it paid to construct tunnels,
cuttings and embankments. It has been said of the English engineers that they defied all natural opposition. In contrast, the
American engineer, with land cheap but labour scarce, avoided
earthworks and built around natural obstacles and may well have
adopted primitive construction methods. Even with rails imported
from Britain, American rail track was built at about one-tenth of
the British cost per mile.

# THE RAILWAY AND DEVELOPMENT DEBATE

Some general aspects of the transport and development debate have been considered in Chapter 1 and the emphasis here is more specifically on railways. The emphasis in much writing on railways has been on their causal role in the development process. For Savage (1959) it 'can hardly be overemphasised, agricultural and industrial development and settlement of the West (of America) would scarcely have been possible without it.' Rostow (1960) saw railways as 'historically the most powerful single initiators of (economic) take off being a main force in the widening of markets and a pre-requisite to expanding the export sector.' Jenks (1944) saw the impact of railways in their demand for equipment and the capital flows generated with consequences for the development of money markets.

Rapid economic growth in the ninteenth-century was certainly associated with an expansion of the supply of transport and this was largely provided by the developing railway networks (BBC, 1993). Railways increased the efficiency and reduced the cost of transport, this being particularly significant for bulk goods in large quantities. In many parts of Africa, railways provided the first real alternative to human porterage and thereby assisted in the elimination of what some considered to be a social evil, a political danger and an economic waste (Harrison Church, 1956). In the Gold Coast (Ghana) the completion of the first rail line in 1903 from the coast at Sekondi inland to the gold mining area of Tarkwa immediately reduced the land transport cost per tonne of imported goods for the mines from £25.35 to about £3.00. The real cost reductions were less obvious where the new railways were in competition with existing water transport (O'Brien, 1983).

The impact was undoubtedly profound. In East Africa the early rail lines were vital to the development of the commercial economy, expanded the range of commercial agriculture and stimulated settlement. In East and Central Africa the 'line of rail' became the zones of economic activity and the rail heads the focal points for settlement expansion and economic input and output (O'Connor, 1965). Mombasa, Nairobi, Nakuru and Kisumu on the Kenya main line provide obvious examples. In Ghana the 'Golden Triangle', Accra–Kumasi–Takoradi, shows a similar concentration of economic activity and Leinbach (1975) has described the significance of the railway in providing the stimulus for the growth of tin mining and rubber cultivation in western Malaysia.

In India, the railways led to innovation in administration, law, education, public health and irrigation and provided a basis for regional specialisation in both agriculture and industry. More land was brought under cultivation and wages and prices rose. In India and other colonial areas while the railways may well have resulted in significant economic changes they must also be seen as instruments of politics, strategy and military control. Many of the African lines were essentially constructed for non-economic reasons and provided the inland penetration as a by-product of the need to demonstrate the effective political control required to justify colonial claims to territory.

While the association of railways with economic development is undeniable there has been increasing evidence to suggest that the relationship was not causal as some earlier writers might have implied nor was the empirical basis for the all-important role of the railways as powerful as was assumed. In the case of Britain it could be argued that it was canal transport which had provided the critical initial stimulus and the railways were simply a response to the inadequacy of the waterways in maintaining the momentum in an established demand. Likewise, in America the railways were responding to, rather than creating demand and they could not in fact compete in price terms with the steamboats. The railways certainly crossed the wilderness, but far from striking out boldly into the wilderness, they really served to link up areas of established economic activity on the east and west coasts previously connected only by a very long sea voyage (Schivelbusch, 1986).

Another possible approach to this issue is to ask what might have happened if the railways had not been developed. In the absence of railways, would there have been more rapid innovation and improvements in the efficiency of road and water transport? (O'Brien, 1983). O'Brien goes on to suggest that nowhere in Europe did railways make the difference between development and stagnation although in countries such as Mexico, Russia and Spain the impact was greater because of the problems associated with the construction of waterways and the limited possibilities for coastal trade. Arguably, by the same token many railways in Africa, Asia and Latin America had relatively greater impact by reducing the impediments to internal and export trade.

It may be suggested that the real significance of railway construction lay in its impact on forward linkages created through regional price differentials for goods and backward linkages into iron and steel industries, engineering and energy industries – the primary rather than the secondary sectors. However, there seems little reason to doubt that at the time the railways were constructed they did serve, by whatever mechanism, to expedite economic growth rates in many, if not most, of the areas in which they were provided.

It has long been recognised (see Chapter 1) that any transport development may have undesirable, negative effects as well as positive, desirable impacts. In the case of India it might be thought that the railways did little more than spread a veneer of modernisation with few constructive and many destructive results. The existing social order was broken down and no genuine process of development took its place. It is possible that alternative policies might have initiated real development – start with narrow-gauge railways and improve as demand built up, keep equipment needs low by interlinked lines so that equipment can be moved to match fluctuations (e.g. seasonal) in demand, low rates to encourage traffic, high rates from ports to interior to discourage imports, raise capital locally, experiment in organisation and train personnel. The strategies were generally otherwise and this served to switch the balance of impact from the possibly more desirable to less desirable end of the scale.

It can be argued, and it has become fashionable in some circles to do so, that no transport system should be considered 'separately from the pressure groups, oppressive economic systems and individually exploitive situations in which they find themselves' (Eliot Hurst, 1973). An essentially Marxist viewpoint would be that railway construction was an instrument for capitalist penetration and while it did effect capitalist development and growth at some localities the overall effect was uneven, debilitating and a contribution to underdevelopment (Pirie, 1982). Thus Slater (1975) sees the construction of railways in East Africa as the means whereby areas of peasant agriculture were brought under German capitalist enterprise to provide income for payment of hut tax and McCall (1977) concluded that colonial transport investment, much of it in railways, was primarily for the promotion of economic, social and political underdevelopment.

According to Pirie (1982), in Southern Africa the 'decivilising' railways contributed to underdevelopment in four ways. First, they induced indebtedness by never being economically viable and being financed by loans that diverted scarce resources away from other uses. Second, they contributed to economic dependency through overseas financial institutions, sources of equipment and supplies and often staffing. Third, they promoted wage labour and proletarianisation and destroyed the peasant economy. Fourth, the railways facilitated labour migration, widened labour supply areas and allowed the persistence of reserves of cheap labour.

It seems reasonable to conclude that railways may at the same time be both the cause and result of economic growth yet their impact clearly transcends the purely economic and embraces social, cultural, demographic, political and strategic considerations. Further, and possibly depending on one's political philosophy, these influences may be seen as either desirable or undesirable and in most cases will anyway be a mixture of the two. To embark on railway construction is not therefore the obvious solution to development problems which some theorists may have implied. Further, the actual impact of any railway construction will vary with the specific conditions prevailing and the organisation and technology adopted.

## THE ORGANISATION OF RAILWAY SYSTEMS

In Britain and America railways were usually initiated, financed and constructed by private enterprise – often involving strong local or regional vested interests based on agriculture, mining, industry or commerce. Operational convenience over time often led to the linking of the individual lines, company mergers and take overs and the creation of larger, regionally based companies with overlapping spheres of influence. Thus in Britain in 1921 over 120 railway companies, many themselves the products of mergers of smaller units, were reorganised into four 'regional' companies – the London Midland and Scottish, the Great Western, the London and North Eastern and the Southern. In 1948 these were nationalised. In North America, the nationalisation apart, the overall process was rather similar. In most other European countries, while initially there were elements of private enterprise, the railways soon emerged as state organisations.

However, even where privately owned, there was considerable state regulation, railways were obliged to publish full tariffs and often had common-carrier obligations which in many cases have only disappeared comparatively recently. In the South American countries, many of them independent for most of the railway era, there was often a strong element of private enterprise and in a number of cases, Argentina and Bolivia being examples, the capital and engineering expertise came largely from Britain. In the nineteenth-century colonies established by Europeans in Africa and Asia it was often the case that the railways were developed, for reasons already noted, by the colonial government and such railways for the most part remain state controlled in the post-independence period.

The railways established in the early colonial period were for reasons of parsimony often built to low design standards and at least cost and in many cases have proved incapable of satisfactorily accommodating the traffic of expanding economies. The railways of Ghana and Nigeria provide good examples of initial considerable impact but are networks on which even the traditional traffic is now handled with great difficulty and where there is no possibility of rail induced economic development taking place. Sierra Leone provides an even more extreme case. Here the government railway was built between 1895 and 1914 and linked the capital, Freetown, with regional centres of Bo, Pendembu and Makeni. The lines were single track of 780 mm gauge, had gradients of 1:50 and numerous sharp curves. The track weight of 14.8 kg per metre allowed axle weights of only five tons. However, these government lines did, despite their deficiencies, stimulate great increases in production – of palm produce in Sierra Leone, cocoa in Ghana, palm produce and groundnuts in Nigeria, tin and rubber in Malaysia – and served to increase local wealth and led to a wide range of other social provision.

It is useful to distinguish a different type of railway provision, that in which the line is developed by a private company specifically for the movement of its own commodities, most frequently for the evacuation of a mineral from deposit to exporting port. Examples would be the line completed in Liberia in 1963 by the Liberian American Swedish Minerals Company (LAMCO) from its iron ore mine at Yekepa to the port of Buchanan and that, also completed in 1963, by the Société Anonyme des Mines de Fer de Mauritanie (MIFERMA) to link Mauritania's western Saharan iron ore deposits with the coast at Nouadhibou (Hilling, 1969b). The *raison d'être* for these lines is the programmed movement of the company's product and any other traffic will be secondary and possibly seen as an impediment to smooth operations and therefore discouraged. Such lines are very much a part of the defined development project and while they may be positive in Gauthier's use of that term (Gauthier, 1970) it does not necessarily follow that there will be stimulation of a wider range of economic activity.

The construction of the LAMCO rail line from Yekepa to Buchanan (Figure 3.1) required the initial provision of air strips to service the construction camps and also access roads. This infrastructure, handed over to the government on the completion of the railway, might have provided for local movements but soon fell into disrepair. The railway was in fact provided with a transit yard at Tropoi which was well located to serve a potentially rich forest zone but the impact has not been great. The railway is not a common-carrier and other traffic is accommodated only when empty ore wagons are available. Nevertheless, by the late 1970s the line was carrying some 60,000 tonnes of timber a year and small quantities of palm produce and rubber and the LAMCO rail bus also carried a small number of non-company passengers (Stanley, 1984). However, he concludes that in a country where road transport was 'abominable' the railway provided the potential for stimulating economic development but this has not been fully realised because of political instability (and more recently prolonged civil war) and an inability to divert attention away from the large-scale mineral and agricultural concessions. Perhaps the government should have taken a more positive line in encouraging LAMCO to provide the necessary capacity and infrastructure to accommodate additional traffic and should then have embarked on development initiatives to exploit the forest resources along the line of rail.

## THE TECHNOLOGY OF RAIL SYSTEMS

Schivelbusch (1986) has argued that a 'perfect road should be smooth, level, hard and straight' and railway lines come as close to this ideal as anything. Yet this triumph of mechanical regularity over natural irregularity, again Schivelbusch's words, is achieved only by considerable engineering and at great cost. Herein lie both the advantages and the disadvantages of the railway as a mode of transport. It is very much an all-or-nothing technology and does not lend itself to piecemeal construction and gradual upgrading – it

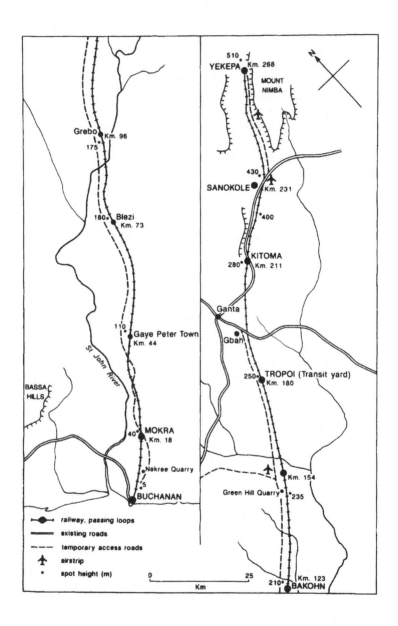

*Figure 3.1* The LAMCO railway, Liberia

*Plate 3.1* In Liberia, the LAMCO railway is of high design standards, allows 9,000-tonnes ore trains and provides passenger and general freight transport.

is not easily modified in the face of changing demand. Certain engineered features of railways, such as ruling gradient, curvature, axle loading and structure gauge, once built can only be modified at very great cost (Nock, 1966).

Perhaps more than with any other mode it is therefore necessary for the planner to make the right decisions at the outset as mistakes could be costly and possibly irreversible at later stages. It is not without significance that rail track is often called 'permanent way' and it follows that, for technical and also economic reasons to be discussed below, railways are not readily adaptable by small increments in capacity in response to fluctuating traffic conditions. Yet engineered to high standards, a rail line can provide a transport link of very considerable capacity, and used properly, transport at very low cost.

The design features of any rail line will be influenced by a variety of interrelated factors (Table 3.1) the principal of which will be the type of traffic, operational characteristics, environmental considerations and materials available. Rail traction is particularly sensitive to changes in gradient and curvature and terrain will be a vital

*Table 3.1*   Traffic loading and support system relations

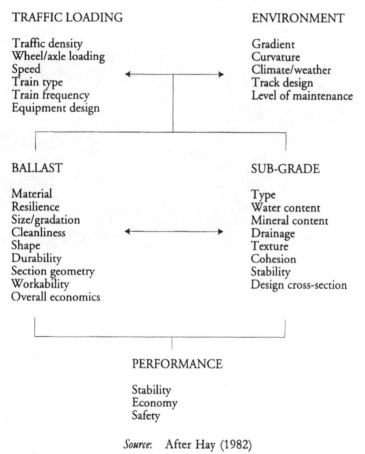

TRAFFIC LOADING

Traffic density
Wheel/axle loading
Speed
Train type
Train frequency
Equipment design

ENVIRONMENT

Gradient
Curvature
Climate/weather
Track design
Level of maintenance

BALLAST

Material
Resilience
Size/gradation
Cleanliness
Shape
Durability
Section geometry
Workability
Overall economics

SUB-GRADE

Type
Water content
Mineral content
Drainage
Texture
Cohesion
Stability
Design cross-section

PERFORMANCE

Stability
Economy
Safety

*Source:*   After Hay (1982)

determinant of route selection. Physical geography is also of significance as an influence on weather and climate, the character of the sub-grade and the availability and suitability of local material for ballast.

## THE RAILWAY TRACK

A number of components combine to support the weight of the loaded train. Basically, the wheel/axle weight is transferred through the rail and the sleepers (cross-ties) and then distributed by the ballast to ensure that the pressure on the sub-grade does not exceed its safe carrying capacity (Clarke, 1957). A standard rail bed cross-section is illustrated in Figure 3.2. The ideal sub-grade will have high internal friction, density and cohesion and low compressibility and capillarity. Load-bearing pressures vary considerably depending on the nature of the material:

| | |
|---|---|
| Alluvium | Up to 7,000 kg per square metre |
| Uncompacted 'made' ground | 7,700 to 23,000 " |
| Soft clay; wet, loose sand | 25,000 to 31,000 " |
| Dry clay; firm sand | 33,000 to 47,000 " |
| Dry gravel soils | 48,000 to 62,000 " |
| Compacted soils | More than 62,000 " |

The ballast holds the track structure to its correct horizontal and vertical line and distributes the weight evenly to the foundations. Good ballast allows effective drainage and the depth required will depend on the axle weights to be carried and the character of the sub-grade. In very dry conditions and natural firmness – the MIFERMA line across the Western Sahara provides an example – high loadings can be supported by shallow ballast. On a rock base 80 mm may be adequate whereas on a wet clay many times that depth may be needed. Average ballast depths are about 0.6 m. Ideally, the pressure on a non-mechanically compacted base should not be more than 0.85 kg per square cm. Ballast depths have to be increased for higher speeds and also more frequent train paths – there is always a tendency for natural recovery after compression and sinking, unless trains follow on too rapidly.

The material used for the ballast is likely to depend on local availability but rock such as granite or limestone crushed to 2 to 4 cm is ideal, being durable, stable and readily cleaned to retain its properties. Crushed slag from steel mills and blast furnaces, prepared pit-run gravels, cinders, burnt clay and sand are all usable but are far from ideal. The frequent passage of trains inevitably causes the ballast to settle and ballast maintenance becomes vital especially on heavily used track.

The ballast anchors the sleepers which support the rails themselves. Sleepers may be of steel, concrete, hard and soft wood and again the choice is likely to be determined by local conditions and availability of materials. Steel sleepers used originally on the MIFERMA line were in seven years reduced by salt corrosion to a flakey mass and were replaced with tropical hard wood. Hard

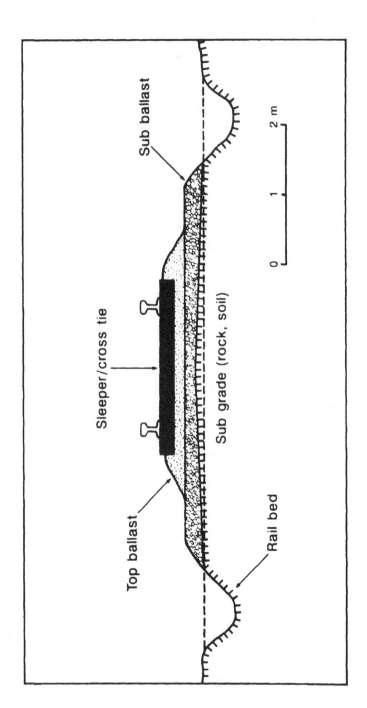

*Figure 3.2* Standard rail-bed cross-section

wood lasts longer but is less amenable to preservative treatment than soft wood. The life expectancy will clearly reflect the material used and maintenance levels but a treated wood can be expected to last between 30 and 50 years with some 50 per cent being replaced after 30 years and 85 per cent after 35 years (Hay, 1982). The spacing of the sleepers will again be a function of their quality, the ballast supporting them, and train weights, and the number of sleepers per kilometre will range from 1,300 to 2,200.

This brings us to the rails themselves. While there are some variations in the actual cross-section of the rail the main differences are in dimensions and the actual weight of steel, usually expressed as pounds per yard or kilograms per metre. Hay (1982) suggests that a crude rule of thumb is that 4.1 kg rail weight per metre will support 1,365 kg of concentrated wheel load. Some of the standard weights are 32 kg (9,500 kg wheel load), 45.5 kg (13,650 kg), 52 kg (15,700 kg) and 60 kg (18,000 kg). The actual cost of the rail is an important element in the capital cost of the track – heavier rail is more durable and reduces maintenance but will only be justified where the loadings are high. However, this is an area in which there is some scope for upgrading as demand increases.

Rail weight is likely to be decided in relation to expected traffic. In China a standard 50 kg rail has been adopted on main lines and 43 kg on less busy routes. In Malaysia the norm is 40 kg on main lines and 30 kg on less used branches. In the mid-1980s the rail weight on the Nanakpur line in Nepal was increased from a low 12.5 to 16 kg and in 1988 the Benin railway in West Africa replaced 22 with 30 kg rail. Part of the Ethiopian railway was in 1987 upgraded to 36 kg to allow axle loads to be increased. It is often the case that the number of sleepers is increased as a part of such upgrading.

## RAIL TRACK GEOMETRY

Suitably dealt with under this heading are the related aspects of gauge, gradient and curvature. Possibly because variations in gauge have such interesting repercussions for the geography of railway operations it is this aspect of railway technology which has received most attention in the non-engineering literature (Best, 1966; Siddall, 1969).

There are, however, two distinct aspects of gauge. The first is the track gauge – the actual distance between the two rails. There is also the question of the loading gauge or the outside width and height of the rolling stock used on a particular track. Thus, whereas Britain and France both have the 1,435 mm track gauge, the French loading gauge is greater and it follows that French rolling stock cannot run through the Channel Tunnel on to British track. Clearly, the track and loading gauges will be a cause and effect of structure dimensions – bridge and tunnel heights and width clearance for example.

Choice of gauge is a fundamental decision in any rail construction project. Initially, as suggested above, many railway lines were one-off links and gauge could be adopted to suit the traffic, terrain and capital available. Problems arise when individual lines join neighbours of different gauge as this means transhipment or changing the wheel bogies, assuming that the different loading gauges can be accommodated. In many national rail systems – India, Argentina and Australia are good examples – rail transport is far more complicated than might appear from a simple map of the lines because the gauges are so varied. In India there were four main gauges and 52 break-of-gauge points (Siddall, 1969) and in Japan the high- speed 'bullet trains' are on the 1,450 mm gauge and the bulk of the rest of the system is 1,067 mm. Argentina has three gauges – 1,676, 1,435 and 1,000 mm.

If gauge is so fundamental, what are the factors influencing choice? Siddall (1969) suggests that they may be grouped under four main heads – existing infrastructure, political factors, technical and economic considerations. In adding to an existing rail system the economic advantages of conformity are considerable. However, political or military factors may sometimes dictate hindering through movement – the adoption of the unique 1,047 mm gauge for the Damascus–Medina line was supposedly for this reason and the Russians saw advantages in a change of gauge between their system and that of the rest of Europe. The choice of the metric 1,000 mm and the Imperial three feet six inches (1,067 mm) by the French and British for their colonies in West Africa may in part have had strategic significance.

Of greater general significance are economic and technical factors, these being closely related. The principal economic consideration will be that of capital cost, a broad gauge line being approximately twice the cost of a narrow gauge. The higher cost derives from the additional earth works and larger structures

*Table 3.2*   Selected rail system features

| | Gauge (mm) | Track weight (kg/m) | Max. axle weight (tonnes) | Min. curve radius (m) | Max. gradient (per cent) | Max. speed (kph) | Traction |
|---|---|---|---|---|---|---|---|
| Nepal | 762 | 16 | 14 | 40 | 1.0 | 32 | S |
| Vicinaux (Zaire) | 600 | 9 | — | — | — | — | D |
| Colombia (National) | 914 | 50–70 | 15 | 80 | 3.8 | 62 | S D |
| Colombia (El Cerrejon) | 1,435 | 62 | 25 | — | 0.3–1.0 | — | D |
| Brazil (Itibara-Tuberao) | 1,000 | 68 | 25 | 202 | 0.3–0.5 | — | D E |
| Carajas-Ponta da Madeira | 1,600 | 68 | 30 | 860 | — | — | D |
| Liberia (LAMCO) | 1,435 | 67 | 30 | 500 | 0.5 | 70 | D |
| Mauritania | 1,435 | 54 | 26 | 1,000 | 0.6–1.3 | 60 | D |
| Tanzania (TAZARA) | 1,067 | | 20 | 20 | 2.0 | — | D |
| (Railways Corp) | 1,000 | 55–80 | 15 | — | 2.2 | — | D |
| Ghana | 1,067 | 29–45 | 16 | 105 | 1.25 | — | D |
| Côte d'Ivoire | 1,000 | 30–36 | 17 | 500 | — | — | D |
| Malaysia | 1,000 | 40–60 | 16 | 142 | 1.0 | — | D E |
| Thailand | 1,000 | | 15 | 180 | 2.6 | — | S D |
| India | 610 | — | — | — | — | — | S |
| | 762 | — | — | — | — | — | S D |
| | 1,000 | 52–60 | 22.5 | — | — | — | S D E |
| | 1,676 | — | — | — | — | — | S D E |

*Source:* *Jane's World Railways*, (1994)
*Notes:* S – Steam, D – Diesel, E – Electric

(tunnels, bridges), the longer sleepers and the increased quantities of ballast. It would clearly be unwise to adopt a gauge greater than that needed to handle expected traffic although the possibility of traffic growth and future line linking should be taken into account. Many of the colonial railways were built to the narrower gauges to reduce cost to a minimum and little thought was given to possible future traffic growth. The 760 mm Sierra Leone railway, the 615 mm Mayoumba and 600 mm Vicicongo lines in Zaire and the general adoption of 1,000 and 1,067 mm in much of Africa provide examples of this.

From a technical point of view the broader gauges allow the use of larger, more powerful locomotives and higher capacity rolling stock for passengers and freight. In economic terms this means more train tonne km per hour at lower cost. There is general agreement that broader gauges provide greater stability (this may be of importance in areas subject to frequent high wind velocities) and allow faster running speeds. However, on well-maintained 1,000 mm gauge track and with suitable rolling stock, speeds of 100 kph are frequently attained.

From the point of view of terrain, the narrow gauge can have much tighter curves and arguably is more appropriate where the the terrain is difficult while on wide, open plains where large radius, high-speed curves are possible the broad gauge is favoured. However, there would seem little agreement amongst railway engineers regarding the ideal gauge and world wide there are nearly 40 different gauges in use ranging from 330 to 1,670 mm with perhaps a dozen in fairly common use (Nock, 1966).

The so-called 'standard' gauge of 1,435 mm would seem to have come about by accident rather than design but is close to the historical width of wheeled road vehicles. The 1,829 mm (six feet) gauge was popular for some early lines in the north-east United States but broad gauge lines are found widely in the very different environments of Ireland, Spain, India, former Soviet Union, Australia and eastern South America. In 1846 a British Act of Parliament standardised on 1,435 mm and this was widely adopted in Europe. Narrow gauges, less than 1,435 mm, are found widely in Africa, Asia and western South America (Table 3.2).

The other important aspects of track geometry are gradient and alignment, especially the number and radii of curves. Gradient is highly significant for railway operators in that additional traction is required on up-grades and brakes have to be used on down-grades both adding to cost of train running. Sudden changes in rise and fall are clearly to be avoided because the effect of gradient on the output of a line is considerable. Table 3.3 shows for a train of 50 wagons of 70 tonnes and a locomotive of 2,000 hp the reduction of output with increasing gradient. It is thought that a one per cent grade is the desirable maximum for high speed passenger operations, 1.5 per cent for heavy freight traffic, 2.0 per cent for high-capacity mixed traffic and 4.0 per cent for low capacity mixed traffic (Busby, 1989). Hence the earlier comments regarding the sensitivity of rail transport to terrain. Lower grades are obtained by cuttings and embankments and are therefore at higher capital cost and there is therefore an important trade-off between these higher costs and the possibility of using larger units or more and faster, lighter units (Hay, 1982).

*Table 3.3*   Track geometry, train speed and output

a) Gradient

| Gradient (per cent) | Speed (mph) | Output (tonne miles per hour) |
|---|---|---|
| 0 | 28.96 | 101,374 |
| 0.5 (1:200) | 9.31 | 32,600 |
| 1.0 (1:100) | 5.30 | 18,540 |
| 1.5 (1:75) | 3.69 | 12,903 |
| 2.0 (1:50) | 2.83 | 9,885 |
| 2.5 (1:25) | 2.29 | 8,009 |

b) Track curvature, train speed and output

| Curvature (degrees) | Speed (mph) | Output (tonne miles per hour) |
|---|---|---|
| 0 | 28.96 | 101,374 |
| 1.0 | 25.54 | 83,384 |
| 2.0 | 22.64 | 79,235 |
| 3.0 | 20.20 | 70,706 |
| 4.0 | 18.16 | 63,546 |
| 5.0 | 16.43 | 57,517 |
| All for trains of 50 wagons of 70 tonnes and 2,000 hp locomotive | | |

*Source:* After De Salvo, 1969

Some curves are almost inevitable but they increase wear on the track, require more maintenance and tractive power. While the effect is not as dramatic as for gradient, increased curvature does serve to reduce output (Table 3.3). To maintain an operating speed of 60 kph, the curves should not be less than 165 m radius, for 100 kph should be 455 m and for 200 kph need to be at least 1,820 m (Busby, 1989).

There are undoubtedly many Developing Countries in which railway operations are now less efficient than they might be simply because of design decisions made in the past, and often for the sake of reducing the capital cost. The Sierra Leone railway, already cited, provides a very good example of this although many other lines of similar design standards are still in operation.

## TRACTION

The form of motive power adopted will depend on the geography of the country (fuel availability, water supplies, electricity generating capacity, terrain), the characteristics of the railway operations (traffic density, length of hauls, numbers of stops, staff, maintenance) and the technical requirements (level of adhesion, power/weight ratios for locomotives, maximum availability of locomotives, ease of operation and maintenance, capacity to deal with overload).

In most countries steam provided the traditional motive power and while steam engines all appear to have immense power, in general it is not an efficient form of traction. Table 3.4 summarises the characteristics of the different forms of traction from which it emerges that electric traction has clear attractions – 'For its out-standing characteristics of availability, suitability for high speed, heavy loads and non-stop runs, robust construction and low main-tenance the electric locomotive is the master card (Majumdar, 1985).

However, the price is high, with considerable fixed installations, catenary systems, electricity sub-stations and the assumption of assured electricity supplies. Only high traffic densities are likely to justify the capital cost involved. The relative cost of fuel for diesel engines and for electricity will be variable depending on location, market situation and means of electricity generation. Indian studies suggest that the costs of the different forms of traction are in the approximate order of steam 100, diesel 95 and electric 65 (Majum-dar, 1985) but diesel has the distinct advantage that it can be phased in with increasing demand and the cost spread over time whereas electricty involves very high initial investment. It is there-fore a question of substituting higher fixed, capital costs for electric traction and the higher, variable running costs of diesel. Fuel alone accounts for some 50 per cent of the running costs of a diesel loco but less than 30 per cent for electric.

In the United States, electric traction was first introduced on some heavily graded routes in 1915 (e.g. Norfolk and Western Railroad) and diesel started to appear in the mid-1920s. The last steam locomotives were built in America in 1953 and in Britain in 1960 but India was still building steam engines in the early 1970s. Diesels were first introduced in India in 1957. It is perhaps not surprising that India and China, each with vast coal supplies, still have very large numbers of steam engines in operation although in both countries electric traction is being introduced on very heavily used routes. In many Developing Countries the general scale of operations and lower capital cost seems to favour the use of diesel traction. In the Sudan in the mid-1980s there was an interesting readoption of steam traction. With the need to move large volumes

*Table 3.4*   Characteristics of alternative rail traction power

|  | *Steam* | *Diesel* | *Electric* |
|---|---|---|---|
| Horsepower Drive | Variable Direct to some wheels | Constant Mechanical Hydraulic Electric | Constant |
| Weight/power ratio | High | Low | Lowest |
| Maximum power | Slow to develop | Instant | Rapid build up |
| Starting trailing load | Limited | High | Very high |
| Crew | Two | One | One |
| Introduction | Slow (fuel, water logistics) | Rapid (no costly line work) | Slow (costly line work) |
| Overload capacity | Poor – speed reduction on grade | Poor – speed reduction on grade | Good – maintains speed on grade |
| Reliability | High | Low | High |
| Adhesion factor | Low | Moderate | High |
| Availability | Low (coaling, water, ash clearing, tube cleaning) | High | High |
| Maintenance costs | Low | Considerable | Low |
| Locomotive life | Long | Short/medium | Long |
| Fuel | Coal, wood, waste from agriculture | Diesel oil | As needed for electricity generation |
| Problems | Fuel availability Crew skills Coaling logistics Fuel consumed when not in use Air pollution Danger of track-side fire | Dependence on one fuel Maximum range about 400 miles | |

*Source:* Based on Majumdar (1985)

of relief food aid the railway was found to be seriously inadequate – after 1959 there had been a gradual change from steam to diesel but lack of spares and inadequate maintenance meant that a large proportion of the locomotives were not operational. A number of steam locomotives were rehabilitated for their advantages of high reliability and low dependence on imported fuel (they could burn local wood) and spares.

## TRAIN PATHS

By train paths is meant the number of trains that can pass along a given stretch of track or through the network in a given period of time. This will be a function of the number of tracks available, on single tracks the number of passing loops, the mix of traffic (speed differences, through trains, stopping trains, passenger trains, freight trains) and the signalling technology. Stops may be planned for commercial (picking up or dropping freight or passengers) or operational (fuelling, water, crew changes) reasons but can also be unplanned (equipment failure, accidents, structural failures).

For any particular track layout it is possible to calculate the theoretical number of train paths. In practice, everywhere, the actual number is likely to be much less, normally by some 50 per cent but in many Developing Countries by 66 to 75 per cent (Nock, 1966). It is worth noting that many of the world's rail routes, and nearly all those in Developing Countries, are only single track and disruptions to the passage of trains, and therefore to the number of train paths, will be unavoidable. A solution once adopted on the MIFERMA line in Mauritania is quite exceptional. There, on an occasion when an iron ore train was derailed, with nothing but flat desert around the derailment site, a bypass track was constructed in a few days and far more rapidly than the derailed wagons could have been removed and the original track repaired! This option is available to few railway operators.

'Up' and 'down' trains on a single track will pass at loops and the number and design of the loops, some variations are shown in Figure 3.3, become critical factors in determining the number of train paths that are possible. Increasing the number of loops is a way of increasing path capacity while increasing the length of loops allows longer, higher capacity trains to be used. It is often the case that passenger or freight stations are conveniently located at passing points. Two examples will illustrate the planning of train paths.

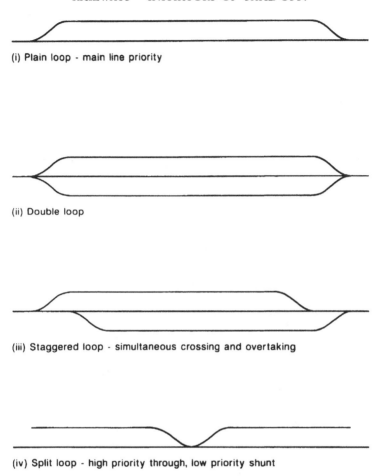

(i) Plain loop - main line priority

(ii) Double loop

(iii) Staggered loop - simultaneous crossing and overtaking

(iv) Split loop - high priority through, low priority shunt

*Figure 3.3*   Typical passing loop designs

In Mauritania, the rail link between the western Saharan iron ore deposits at F'Derik and Tazadit and the exporting port at Nouadhibou is across 650 km of harsh desert, part reg, part barchan, part extensive linear dunes. In these conditions it was decided to maximise train size (135 mineral wagons of 75-tonne capacity each) and minimise train paths so that the fewest possible trains were crossing the desert at any time. This is represented in Figure 3.4, from which

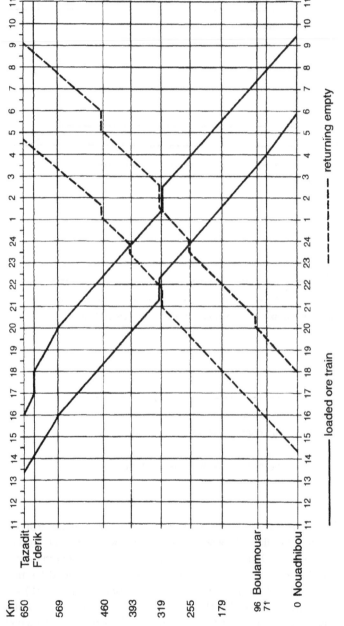

*Figure 3.4* Nouadhibou–Tazadit line, Mauritania, train paths
*Source:* From MIFERMA Information (1965)

loaded ore train    - - - - - returning empty

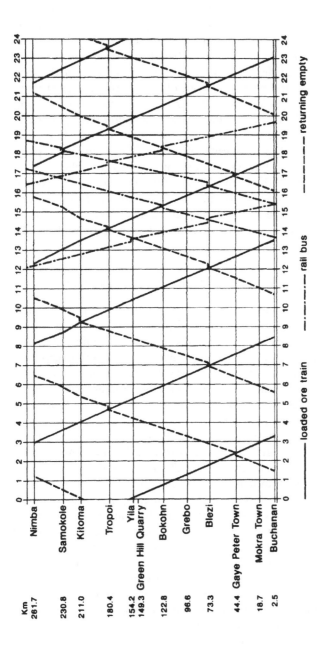

*Figure 3.5* Buchanan–Nimba line. Liberia, train paths
*Source:* From Koenen (1969)

it can be seen that in each 24 hours there are two loaded trains down to the coast and two empty trains making the return journey. There are eight possible passing loops with three (Km 255, Km 319 and Km 393) providing the normal scheduled passing points.

In the case of the LAMCO mineral line in Liberia the distance is only 262 km and with train control more easily achieved than in Mauritania it was decided to use trains of 85 wagons of 90 tonnes each but with five train paths in each direction daily (Figure 3.5). The situation is complicated slightly by the introduction of two passenger rail buses in each direction. The number of passing loops available is nine and all except two provide scheduled train crossing points (Koenen, 1969). Two of the loops are more elaborate and provide freight transit facilities but not passenger interchanges. With trains running at relatively high speeds for their weight, 70 kph, control is by means of a computerised signalling system.

## THE CAPACITY OF RAIL SYSTEMS

The capacity of rail systems is a function of two principal determinants – the maximum size of the train as dictated by track geometry and technology and the number of train paths. Nash (1985) suggests that for a single track there can only exceptionally be more than two trains each way per hour and for a suitably signalled double track 20 to 30 is possible. Sophisticated signalling systems can allow headways of as little as 90 seconds for trains travelling at over 150 kph.

Returning to the specific examples already cited, operating for five days a week and 50 weeks a year the Liberian LAMCO railway has a theoretical capacity of 9.4 million tonnes a year while that of MIFERMA in Mauritania has 5.0 million tonnes. In practice both have carried well in excess of these figures by increasing train paths or adding wagons and locomotives. In contrast to these well-designed and efficiently engineered company owned lines, the narrow-gauge, minimal-design standard Sierra Leone railway dating from the colonial period had very low operating speeds (average 24 kph) and very low capacity and was closed in 1969, being unable to compete in terms of service with road transport. In general, any well designed, properly engineered and well maintained railway line allows fast, scheduled, frequent movements and gives a high capacity at low unit cost. However, if the railway is to be economically viable the throughput must be maintained at high levels.

## RAILWAY ECONOMICS

Only a brief outline of the principal factors can here be considered although railways have attracted a great deal of attention from economists. Perhaps because they were very large operations with special problems arising from the considerable element of 'sunk' costs, they display problems of allocating the total costs to a wide variety of outputs which share the facilities – this will be especially true of the general purpose government railways but not for the single purpose company lines. The short-run marginal costs are thought to be very low (Button and Pitfield, 1985) and maximum efficiency may require monopoly conditions. Because of the all-or-nothing nature of railway systems already noted, railway investment is very 'lumpy' and the mode does not lend itself to neat incremental adjustments of capacity to demand.

There has been considerable debate regarding the proportions of fixed (indirect) and variable (direct) costs in the total cost of rail operations (Joy, 1971; Hay, 1982). Although individual estimates of fixed costs vary from 25 to 90 per cent, with time span adopted as a main factor influencing the outcome (Hay, 1982), there is a general concensus that a persisting influence of large elements of capital cost sunk at times long past and the considerable costs associated with maintaining the track, structures and equipment provide railway managers with particular problems. Costs related to fuel, manning and locomotive maintenance are almost entirely a function of throughput while some 80 per cent of the costs associated with rails, sleepers and ballast also derive from traffic levels. However, only about 30 per cent of the cost of structures (bridges, culverts) is variable with throughput and as little as five per cent of signalling and communications.

It has been suggested (Waters, 1985) that of greater significance than fixed costs is the identification of the avoidable costs, those costs that can be saved by reducing capacity, and the incremental costs that will be incurred by expanding throughput. Also of importance to the decision maker will be the opportunity costs, the costs of adopting one strategy rather than another. It will be clear from the above discussion of technology that in theory there is almost infinite scope for substitution between technologies, and therefore costs, although in practice the choices may be reduced by past decision making. Technological choice is often on a line-by-line basis (reduce capacity? increase capacity? replace equipment?) rather than system basis and financial criteria dominate the decision making process now as they did in the past.

A well-established feature of railway operations is the broadly constant returns to scale (Dodgson, 1985) and marked decline of unit costs with increasing throughput which allows the spreading of the fixed costs. As Figure 3.6 indicates, with increasing traffic unit costs fall until such time (B) as congestion, delay, shortages of locomotives or stock lead to increasing costs. If capital is then invested for improvement (C–D) the unit cost will rise and only fall again if traffic volume goes on increasing. There is clearly a danger

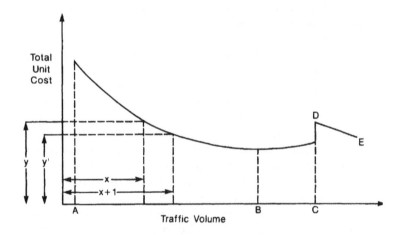

A - B  Declining unit cost as overheads are shared by increasing units of traffic

B - C  Increasing volume causes congestion, delays shortages of rolling stock/locomotives. Unit costs rise

C - D  Cost of additional capacity increases unit costs

D - E  Additions to traffic reduce unit costs again

*Figure 3.6*   Rail traffic and unit costs
*Source:*   After Dodgson (1985)

100

in this situation in which improvements are effected and capacity increased but the traffic does not go on rising. The diagram also illustrates what some economists think should be the basis of pricing. The addition of one unit to traffic (x) reduces unit cost from y to $y^1$ and it is this incremental cost rather than full distributed cost which should be passed to the user.

It has been suggested (Fowler, 1979) that the longevity of railway assets and the complexity of the joint operations dissuaded operators from measuring the cost of moving each passenger or each tonne of freight and the value of goods was often adopted as the main guide to pricing – what the traffic will bear. Yet this is not really satisfactory and the railway operators in Developing Countries should make some attempt to determine costs so that effective decisions can be made about the relative costs for different modes. Only in this way can resources be utilised rationally and transport investments allocated effectively.

## MAINTENANCE

Maintenance is often inadequate in many Developing Countries yet for efficient railway operations is essential. A maintenance cost curve is given in Figure 3.7. With the construction of a new line or rehabilitation of an existing line (A) the maintenance costs will be reduced. There will then be a period of time (B to C) when costs bottom out and may be considered as normal. With increasing

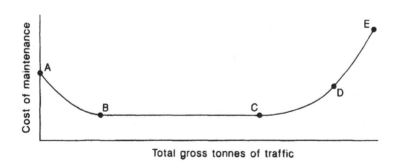

*Figure 3.7*  Rail traffic and maintenance costs
*Source:*  After Hay (1982)

101

traffic, rails and sleepers wear, ballast fails and horizontal and vertical track line is lost and this will require increased maintenance and cause a rise in costs (D). If this maintenance is not carried out and is deferred, an all too common feature in many Developing Countries, the eventual cost of maintenance will be much higher (E). The moral is quite simple – Developing Countries will find it easier and cheaper to keep up with maintenance than to catch up with it. In his 1988 budget speech, President Babangida of Nigeria called for a 'maintenance culture' to replace the 'replacement syndrome' and his advice could well be followed in many other Developing Countries and it is possibly a fault of the financial aid agencies that they are more ready to lend for new projects than they are to ensure that schemes have built-in resources for sensible maintenance.

## RAIL NETWORKS

There have been numeous studies conceptualising and describing the evolution of rail networks (Meinig, 1962; Taaffe *et al.*, 1963; Wallace, 1958; Quastler, 1978) and numerous methods of analysing network characteristics (Haggett and Chorley, 1969). Some of the features of rail networks are particularly significant for the manner in which they allow railways to influence development relative to other modes.

In network analysis it is usual to identify paths, trees and circuits and in railway terms these might be equated to single isolated lines, lines with lateral branches and systems with built-in circuits. In this hierarchy of network types many railways in Developing Countries are of the simpler kind and in evolutionary terms many of the railways did not develop beyond the stage of providing *localised linkage* even sometimes only isolated and complementary to existing modes (e.g. rail bypasses to river rapids along the Zaire river) or simple penetration routes inland from ports, as were many of the early colonial lines in Africa and some of the more recent mineral lines. Liberia's railways were of this type (Figure 3.8).

In places a degree of integration has taken place. Ghana's lines inland from the coast at Accra and Takoradi met at Kumasi but in the 1960s were joined by a cross link further south. In Taiwan, between 1979 and 1982, two previously separate networks were linked and the gauge unified. *Intensification* produces an increasing

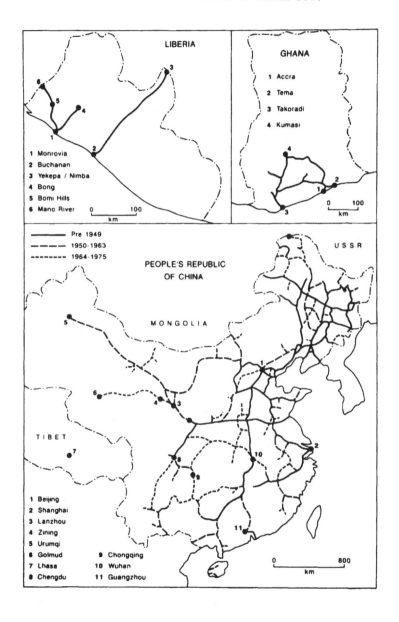

*Figure 3.8* Selected rail networks

number of circuits sometimes with feeder lines into peripheral regions. China is perhaps the best example of this process (Figure 3.8) and one of the few Developing Countries in which there has been large-scale railway construction in recent years.

In India, some low-capacity branches have been closed and this process of *selection* can become more marked, especially where road transport becomes more intensive and provides effective competition – not too difficult given the nature of railway economics and the network features here being described. The three significant measures of network efficiency are density (route miles in relation to area), connectivity (the extent to which places have direct connection with other places) and accessibility (the number of lines needed to get from one place to another). On these counts, and certainly in contrast with road systems, rail networks in most Developing Countries are of very low density, low connectivity and limited accessibility. It is also the case that access to the network (for railways this is at stations and sidings) is very limited whereas for roads it is virtually infinite (see Chapter 5). It follows that the spatial impact of railways may also be limited and a possible effect, especially where the network never developed beyond the simple branching form, may be the concentration of economic development at a relatively small number of nodes or along fairly narrow corridors. Railway networks of the simpler type may well reinforce trends towards regional inequality and while they may be an efficient and cost effective way of moving bulk products to exporting ports they do not provide for the wide distribution of the range of imported goods nor for domestic trade and circulation. Many of the African rail lines traverse extensive areas which produce little or no traffic and where the railway has had no apparent impact on local development.

## RAILWAY INVESTMENT

From the above descriptions of the technical, economic and network characteristics of rail transport it would seem clear that the mode is best suited to specific, well-defined transport functions and is rarely ideal for multi-functional operations. This is not to say that railway systems cannot have a broad-based role in Developing Countries – India and China provide examples where they do just this – and in origin many of them were established to serve a variety of economic, social, political and integrative roles. Yet it is also clear that many of the railway systems in the developing world are struggling with vast backlogs of maintenance, stiff competition from road haulage and declining traffic – Ghana, Nigeria, Senegal and Uruguay are good examples .

Railway investment in Developing Countries is likely to be used for one of three main purposes: first, for the construction of new lines, second for increasing capacity that is reaching its limits and third, to improve the quality of service or rationalise operations on loss-making lines. One of the most-used words in the development literature today is 'rehabilitation' and in many Developing Countries this is urgently needed for their rail networks and is either planned or already in progress.

In the first category, new lines may either be extensions to existing networks or one-off links. Neither type is likely to be justified economically unless there is a clearly identified and assured high-volume traffic, ideally a bulk mineral, but also possibly forest or agricultural products in large quantities, although the latter can often have marked seasonality and fluctuating demand for which railways are not ideally suited. It is more than likely that any such lines will in fact be associated with a specific development project such as the opening of a new mine. The Liberian and Mauritanian lines already described were of this type. Where the rail line is part of a wider development scheme there is likely to be the distinct advantage that the risk capital will come from the mining company but the possible disadvantage that the line will have more restricted general development potential and introduces the undesirable influence of the multinational corporation and new elements of dependency.

In Colombia, the newly opened 380-sq-km high-quality bituminous coal deposit at Cerrejon started exporting in 1987 by way of a new 150-km rail link with the coal exporting port facility at Puerto Bolivar. In Brazil, work started in 1978 on a 890-km line to link the Carajas iron ore deposit with the port of Sâo Luiz. In 1984, some agricultural produce that had previously been moved by boat on the Rio Pindare was transferred to the railway and transit times and costs reduced by 90 and 30 per cent respectively. Agricultural collection stations were established at Alto Alegre (Km 265) and San Pedro de Agua Blanca (Km 650) and in addition to the three 15,000- tonne ore trains daily there are also one general freight and two passenger trains per week. The line traverses an area of rich forest with great agricultural and mineral resources and there is potential for the integration of road and river transport with the new rail line. Road vehicles are already being carried piggy-back style and there are plans for multi-modal regional transhipment centres. There are longer-term plans for electrifying the line and in this case there is undoubtedly the traffic which would justify the cost of such a change.

In Gabon the long-planned scheme to link large iron deposits at Belinga in the north-east of the country with the port of Owendo, near Libreville, failed to attract World Bank support on the grounds that the only beneficiary would have been the single, private mining company. In 1974 the Gabon government decided to go ahead without World Bank support but at Booué the line was diverted south-eastwards to existing manganese mines at Moanda. This meant that the two million tonnes of manganese previously exported by way of a 76-km aerial ropeway to M'Binda and the Congo railway to Pointe Noire could in theory be shipped out through Gabon itself. The line was completed in 1987 and attracted considerable export timber traffic, some general import goods and also passengers. The operation faced mounting deficits attributable to the high investment costs, organisational expenses and a lack of balance between the costs and the tariffs – it is doubtful if the traffic could bear the costs that would be necessary to cover all expenses.

While the economic viability of the Cerrejon and Carajas lines is not in doubt the Gabon line looks uncertain. The possible future construction of the extension to the iron ore deposits at Belinga could assure its financial position and there are signs that timber production and export could increase now that there is improved access to large forest areas – this could be only a short-term advantage unless sustainable exploitation is encouraged and of that there seems little sign at present.

In China after 1949 (Figure 3.8) some 1,000 km of new line were added each year (Rongju, 1984) and in the 1980s new lines were largely prompted by a 10-year programme to increase coal carrying capacity to 600 million tonnes a year. Eight new lines are being built, including a 379-km electrified, double-track link from the Shanxi coal producing centre at Datong to Beijing. In China most new lines would seem to be for specific bulk traffic and the linking of existing or new production centres with the main areas of consumption. However, some lines must be explained in political terms (Leung, 1980) and as an attempt to incorporate geopolitically sensitive peripheral areas (Hsinchiang, Yunnan, Kuangshi, Fuchen) and provide greater inter-regional balance and spatial integration.

Much of China provides exceedingly difficult terrain for railway construction, the distances are great and large areas generate little traffic. Work on the 2,200-km line from Xining to Lhasa (Figure 3.8) was stopped at Golmud because of terrain difficulties, seismic disturbances, low temperatures and need to pressurise passenger cars because of the altitude. This, and other lines such as those from Chengdu to Kunming, Lanzhou to Urumqi and Yington to Xiamen, have involved massive earth moving and recent rail projects in China have involved the construction of some 4,000 tunnels, totalling 1,800 km, and 14,000 bridges with a total length of 1,000 km (Rongju, 1984). China must undoubtedly be seen as the main frontier in present day railway engineering and an area where much construction has still to be undertaken.

There are some countries – Niger, Chad and Central African Republic being the most obvious examples – which may well be bypassed completely by the railway age. Although in each case there have been proposals to extend the railways from neighbouring territories, it is difficult to identify the traffic which would justify the investment involved.

In many countries there has been investment in upgrading projects. Much of the recent railway engineering in China has been to increase capacity on existing lines. Much of the system is overloaded and the 20,000 freight trains running daily now carry only 70 per cent of the freight to be moved (*Lloyd's List*, 8/3/1989). Single tracks are being doubled and heavily used lines electrified.

India has one of the densest rail networks anywhere and is the world's second largest system under unitary management. The system employs nearly 1.5 million people and accounts for two-thirds of all the freight moved (well over 300 million tonnes a year) and at any time there will be some one million Indians travelling by train – many of them on the outside as well as inside the cars! Not surprisingly, there is considerable pressure on the multi-gauge system much of which is of considerable antiquity and was developed in a very piecemeal manner. Some 75 per cent of the freight and 55 per cent of the passengers are moved on the broad gauge lines but gauge conversion is taking place where the narrow gauge lines are at full capacity. Elsewhere, the narrow gauge tracks are being equipped with heavier rails for 12 to 14 tonnes axle loads. By 1987, 7,529 km of track had been electrified (Delhi–Calcutta, Delhi–Bombay) and there are plans to phase out steam traction by the year 2000. Improvement of signalling is in progress on main lines. Freight capacity is being increased by greater use of permanently coupled 'block' trains and gross weights on some lines have been increased to 4,500 tonnes with experimental running at 9,000 tonnes. Container movements number over 40,000 units a year and there are connections between ports and Inland Container Depots at Bangalore, Guntur, Anapati, Coimbatore, New Delhi, Amingao, Dhandan Kalon and Tughlahabad with computerised freight control. However, re-equipment plans have been slowed down in the face of inflation and foreign exchange problems.

India, more than most Developing Countries, and as a result of the scale of its railway operations, has the advantage of having developed a considerable range of railway-associated industries. In

*Plate 3.2* India has a dense rail network, heavily used lines have been electrified and the railways contribute, as in Bombay, to intra-urban as well as inter-urban transport.

particular, India has locomotive, freight and passenger car construction works with an emphasis on low-cost, unsophisticated and solid products suitable to the conditions in the country but also for other Developing Countries and has become a considerable exporter of rail hardware and also expertise. Rail India Technical and Economic Services (RITES) has provided railway consultancy and management services in many other Developing Countries including Algeria, Ghana, Iraq, Jordan, Nigeria, Philippines and Zimbabwe.

In the late 1980s many countries were trying to increase rail capacity. Algeria was adding to the number of passing loops on single tracks and increasing loop length from 800 to 1,200 m. In Cameroon there has been realignment in difficult terrain between Douala and Yaounde with new tunnels and gradients that allow 10,800-tonne trains. In Congo, 16 million cubic metres of earth works have been undertaken and three tunnels and 19 bridges constructed to ease gradients from 2.7 to 1.5 per cent on the Congo–Ocean railway where it rises from the coast to the continental plateau. With curve radii minima increased from 90 to 300 m running speeds have been increased from 60 to 80 kph. In Egypt a five-year development plan for the railways initiated in 1987 brought double tracking to the Sohag to Qena and Dekenios to Malarya lines and upgrading on several other routes. On Ghana's Takoradi to Kumasi line, rail weights have been increased in sections and there has been some realignment and increase in curve radii and re-ballasting in an attempt to reduce all too frequent derailments on a line which carries the bulk of the country's mineral and timber exports. In 1989 Nigeria embarked on a programme of track strengthening and curve elimination.

In Iran, the war with Iraq slowed down work on a scheme to create 10,000 km of new track. The Bandar Khomeini to Tehran line was double tracked but proposed electrification was shelved. Kenya has been installing heavier rails and intends to double track from Mombasa to Nairobi. The Damman–Riyadh line in Saudi Arabia has been realigned and partly double tracked for higher-speed operations while in Turkey higher-capacity wagons have been introduced and priority given to double tracking and upgrading of the signalling on the 577-km iron ore transit route from Divurgi to Adana and Iskenderun.

On many more lines shortage of foreign exchange and neglect of maintenance has made rehabilitation a first priority. Bangladesh received World Bank assistance for rehabilitation of the track from Parbatipur to the the river port of Khulna and in Egypt the lines from Alexandria to Mersa Matruh and from Helwan to Baharin have been brought back to a satisfactory standard. Likewise Ghana's Eastern and Central lines. Perhaps more surprisingly there has been need for a $157 million rehabilitation programme for the TAZARA (Tanzania–Zambia Railway) line completed by the Chinese as recently as 1976. With a design capacity of five million tonnes a year the actual throughput never exceeded 1.5 million as a result of serious problems with locomotives, track and rolling stock quality and maintenance. The locomotives and rolling stock are being replaced long before normal life expectancy has been reached. In 1988 Uruguay embarked on railway rehabilitation after the government forced closure of one-third of the network and reduced staff by 70 per cent.

It would be wrong to imply that great capital investment is necessarily needed to improve railway capacity. In all forms of transport the first priority must be to ensure that the existing capacity is used to maximum efficiency and its fullest potential before money is spent on replacement, elaboration or extension. There is certainly considerable evidence to suggest that many of the railways in Developing Countries, albeit they started with inadequacies of design, have not been well managed and properly maintained.

## POLICY IMPLICATIONS

Many Developing Countries have inherited rail systems built cheaply and to design standards which are far from adequate by any present day criteria; limit their capacity and increase their costs. When originally built, the railways often had a virtual monopoly on mechanised transport but this has been steadily eroded especially by road transport which in many places provides a form of transport better suited to the needs of Developing Countries and arguably does not always have to pay its full costs. Lack of capital and foreign exchange has limited essential investment and maintenance so that the efficiency and capacity of rail systems – and hence their ability to compete – has been further reduced. Declining traffic means declining revenue and with large throughputs necessary to cover operating and maintenance costs the downward spiral continues.

Developing Country railway systems are for the most part

government owned and therefore become a considerable drain on central government resources and in some cases are being subsidised to the extent of 300 to 400 per cent of their revenue. In competition for government finance, inefficient railway systems stand little chance. In this as in other areas some countries are looking to private-sector involvement as a solution to the problem. Côte d'Ivoire has already moved in this direction and early in 1995 legislation was passed by the Mexican Congress which will allow private investors to operate full rail services on a concession basis which will also allow for the construction of new lines. Mexico's 26,000-km network has an average train speed of 25 kph, half that of a container ship and less than one-third that of a lorry. Although container traffic has been increasing much of the cargo spends a great deal of time sat in transit yards and the railways move only a small proportion of the general cargo. Tendering for concessions commenced late in 1995 and it is thought that there will be considerable private interest in some of the longer, more heavily used lines.

In many Developing Countries the railways are heavily overstaffed in relation to throughput and while this may be justified on social grounds it does little to help operating finances. Most railway operations are capital-intensive and there is only limited scope for utilising low-cost labour. China provides an exception and there even major earthworks are undertaken by labour-intensive methods. In Burkino Faso an attempt to 'mobilise the people' to construct the Ouagadougou to Tambao extension seems to have made no progress.

Because they are capital-intensive, railways will be economically viable only where they are used extensively and there are only certain market conditions which will satisfy this requirement. It has already emerged from the above discussion that large volumes of bulk freight (fuels, ores, forest products, agricultural products where not too seasonal) provide the best potential for rail transport except where water transport is available or where pipelines can be used. Such freight will be the most likely basis for further railway development but the movement of containers in large numbers over long distances and also the movement of large numbers of passengers over short to medium distances, possibly in an urban context (Bombay comes to mind), also provide suitable markets.

However, large parts of the developing world have low traffic-generating potential, the distances are often great, the demand slight and highly dispersed and the possibility of rail revenue covering costs is virtually negligible. Governments have seriously to consider whether in these conditions rail transport is justifiable although there may well be non-economic arguments to be taken into account. Environmental considerations may not yet loom large in the equation in most Developing Countries but should not be excluded from the reckoning. Where it is decided that rail transport is the right mode then it must be provided with the technology appropriate to the task and must be managed and maintained effectively. In the mid-1980s the World Bank, with over $5,000 million in loans for railway projects, tightened its lending conditions because of the poor performance of so many of the projects it had supported and called for greater selectivity, sharper focus and more realistic expectations in project appraisal.

The network, operating and economic characteristics of rail transport make it an inflexible mode, lacking adaptability and ability to respond to changing market conditions. The possible role of railways as a factor influencing wider aspects of development can only be assessed in these terms and in consequence is often likely to be limited. Where the conditions are right and the whole properly managed, rail transport can still be a powerful factor and as roads become ever more heavily congested must certainly not be excluded from consideration.

# 4

# AIR TRANSPORT – THE HIGH-COST SOLUTION?

From the first airport for Zeppelin airships in 1910 (Taneja, 1989) and the first scheduled passenger service by Curtiss flying boat between St Petersburg and Tampa, Florida, in 1914 (Vance, 1986) air transport has experienced remarkable technological transformation and phenomenal growth, especially in the years since the Second World War. There is now the paradox that air transport is seen as both the most sophisticated of transport modes and in many parts of the world the provider of pioneer mechanised transport. Discounting primitive forms of human and animal freight movement there are many areas where it is the case of air transport or no transport. It has been said that Colombia went from the mule to the aeroplane in one move.

The initial development of air transport was most rapid in Europe but after 1925 America assumed a leading role, a position it has maintained to the present day. Early operations were costly but the obvious advantages of speed for the movement of mail had new airlines competing frantically for government mail contracts without which they would never have been economically viable. In 1936 the introduction of the Douglas company's twin-engine, non-pressurised DC3 with a maximum speed of 175 mph (280 kph) and range of 500 miles (800 km) gave a capacity, speed, productivity and reduced operating cost which released passenger carriage from the inhibiting need of mail contract subsidy. Over 10,000 of these dependable work horses were produced and sixty years later some hundreds are still in regular use. In many Developing Countries they initiated domestic air transport (e.g. Ghana, Mauritania) and are still the mainstay of Colombian services to many remoter settlements. The DC3 allowed the widening of the geographical pattern of air transport operations.

During the 1920s and 1930s most world regions took their first steps towards developing air transport. In 1919 the British Royal Air Force surveyed the Cairo–Cape route and in 1920 a Britain to South Africa flight was completed with 21 stops, 110 hours in the air and a total journey time of 45 days – far longer than the sea journey! When a regular through service was started in 1932 the journey was reduced to 10.5 days and by 1937 with the use of flying boats this had come down to 4.5 days. The 1920s also saw extensions of Imperial Airways in India and in 1945 Tata Airlines was created with 12 American ex-military DC3 aircraft; a year later this was to become Air India. The network pattern was still essentially that of chains of airports on imperial routes. In Mexico, American interests started air operations in the 1920s especially to service expanding mineral and oil field exploitation and with German influence there was expansion of services in South America. In Colombia, the Barranquilla to Bogota surface journey of a week was reduced to six hours by air.

Improvements in aircraft design during the Second World War were to have a profound effect. The unpressurised DC3, with a ceiling of around 3,300 m, was never able to get out of the lower, most turbulent layer of the atmosphere. The introduction of the pressurised B29 Superfortress with a range of over 11,000 km was to pave the way for the post-war Stratocruiser and raised the range and capacity of air transport.

As a means of propulsion the propeller is most efficient at about 480 kph but as speeds increase the aerodynamic properties are lost. For higher speeds an alternative was needed and this was to be the jet engine in which the exhaust provides the propulsive force. Work on the jet engine during the 1939–45 war led in the 1950s to aircraft such as the Comet, Caravelle, Douglas DC8 and Boeing 707 which were to revolutionise air transport – long-haul air transport came of age.

For speeds above that at which the propeller is most efficient but below that at which the jet really comes into its own, the turbo-propeller aircraft combined the qualities of the two by using the thrust of a turbine in part to power a propeller. This is an efficient and also economical form of propulsion for smaller to intermediate size aircraft at speeds up to about 725 kph and turbo-propeller aircraft such as the Electra, Viscount and Fokker Friendship became common on shorter and medium hauls. The introduction of the turbo-fan or bypass jet, in which exhaust thrust is augmented by the bypass air, greatly enhanced the power output and also had the effect of reducing noice. This was the power unit which made possible the wide-bodied aircraft such as the 747, DC10, TriStar and Airbus from 1970 onwards.

Some of the characteristics of selected aircraft are presented in Table 4.1. The improvements in technology were to have profound implications both in the air and on the ground. The range and capacity of air transport were vastly increased, the real cost declined dramatically, and air transport in the 1960s and early 1970s was expanding at 14 to 15 per cent a year. On the ground the larger aircraft needed much longer runways and more extensive servicing areas while the passengers and freight they carried called for ever larger terminal facilities and improved ground access to airports.

## REGIONAL CONTRASTS IN AIR TRANSPORT

The exceptional rate of technological change was matched by unrivalled expansion of air transport (Doganis, 1985) but this expansion was far from evenly spread. The regional traffic statistics provided by the International Civil Aviation Organisation (ICAO) are not wholly satisfactory in that a region such as Asia and Pacific includes countries at many levels of development. Figure 4.1 does not represent the totality of air transport because it is based on the scheduled services of the airlines of ICAO member countries and ignores the non-scheduled operations which are of particular importance in many Developing Countries. Nevertheless, it shows the overwhelming dominance of North America and Europe in global air transport.

However, since 1971 the region with the most consistent growth has undoubtedly been Asia and Pacific, including some of the high growth economies of the Pacific rim – Japan, South Korea, Taiwan, Hong Kong and Singapore in particular. Indeed, growth in this area has resulted in some erosion of the dominance of North America and Europe and was based on expansion of international and domestic traffic. In North America and Europe the domestic element has shown only slight growth, presumably because air transport in these areas is in competition with highly developed surface transport, and the international traffic growth has been only modest. In Latin America and Caribbean the growth of air transport has also been modest while in Africa, with a large

Table 4.1 Selected aircraft characteristics

| | Type | Introd. | Power | Pass. | Cruise speed (kph) | Max. take-off weight (kg) | Max. payload range (km) | Take-off run (m) |
|---|---|---|---|---|---|---|---|---|
| a) Airliners | | | | | | | | |
| Douglas DC3 | S | 1934 | P(2) | 32 | 266 | 11,430 | 2,430 | 1,000 |
| Boeing 707 | L | 1958 | TJ(4) | 179 | 840 | 148,500 | 9,600 | 3,050 |
| Douglas DC8 | L | 1958 | TJ(4) | 189 | 880 | 143,000 | 7,550 | 2,815 |
| HS 748 | S | 1962 | TP(2) | 48 | 430 | 20,000 | 2,996 | 1,130 |
| Fokker F27 | S | 1967 | TP(2) | 52 | 470 | 20,500 | 1,805 | 1,050 |
| Boeing 747 | L | 1970 | TF(4) | 490 | 970 | 351,000 | 11,600 | 3,050 |
| DC10 | M/L | 1971 | TF(3) | 332 | 900 | 195,000 | 7,400 | 3,170 |
| De Haviland Dash 7 | S | 1977 | TP(4) | 50 | 428 | 21,300 | | 690 |
| BAe 146 | S | 1981 | TF(4) | 93 | 700 | 42,000 | 1,700 | 1,220 |
| Embraer Bandeirante | S | 1983 | TP(2) | 30 | 335 | 5,900 | 2,000 | 675 |
| McD D MD-87 | S/M | 1987 | TF(2) | 139 | 840 | 63,500 | 4,400 | 1,910 |

## b) Utility aircraft

| | | | | | | | | |
|---|---|---|---|---|---|---|---|---|
| Beechcraft Baron | Util. | 1958 | P(2) | 6 | 362 | 3,900 | 1,700 | 610 |
| Beechcraft Bonanza | Util. | 1959 | P(1) | 4 | 319 | 1,542 | 1,648 | 305 |
| Pilatus Porter | Util. | 1959 | TP(1) | 10 | 213 | 2,800 | 730 | 197 |
| Short Skyvan | Pass/Fr | 1963 | TP(2) | 22 | 352 | 10,400 | 876 | 1,042 |
| De Haviland Twin Otter | Feeder | 1965 | TP(2) | 20 | 338 | 5,670 | 1,200 | 213 |
| B-N Islander | Util. | 1966 | P(2) | 9 | 257 | 2,857 | 1,154 | 270 |
| B-N Trilander | Util. | 1971 | P(3) | 17 | 280 | 4,240 | 1,130 | 255 |
| Embraer Ipan | Agric. | 1971 | P(1) | — | 212 | 1,800 | 938 | 200 |
| Embraer Xingu | Util. | 1976 | TP(2) | 10 | 389 | 5,600 | 2,410 | 520 |
| Short Super Sherpa | Util. | 1984 | TP(2) | 30 | 298 | 11,600 | 740 | 1,000 |
| Aviocar C-212 | Util. | 1984 | TP(2) | 26 | 300 | 7,700 | 1,440 | 610 |
| Pilatus PC-12 | Util. | 1989 | TP(1) | 4 | 497 | 4,000 | 2,966 | 310 |

*Notes:* S/M/L – Short/medium/long range
P – Piston
TP – Turbo-propeller
TJ – Turbo jet
TF – Turbo-fan jet
(2) – number of engines

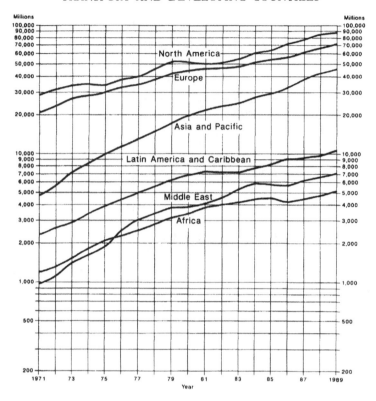

*Figure 4.1*   Regional growth in air transport
*Source*: ICAO reports

number of low-income countries, the growth level has been the lowest of all regions. It is anticipated that globally air transport will increase at around 6.7 per cent a year until the year 2,000, with routes serving Asia showing well above average growth rates (Figure 4.2).

The regional concentration of air traffic is also shown in the ranking of activity by country (Table 4.2). In terms of total passenger kilometres the rank order reflects factors such as country size (USA, Russia, Canada, Australia, Brazil), significance of domestic air transport (USA, Russia, Brazil), historical and colonial legacies (UK, France) and locational advantage (Singapore). It is interesting that China is the only Developing Country to feature in

# AIR TRANSPORT – THE HIGH-COST SOLUTION?

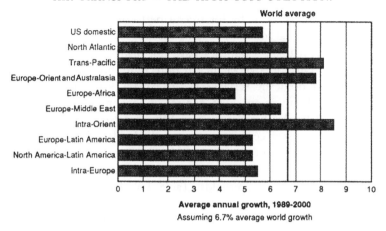

*Figure 4.2* Predicted growth in air transport
*Source*: ICAO reports

*Table 4.2* Air passenger transport – rank order by country, 1992

|  |  | Total passenger-km (millions) | International passenger-km (millions) | Domestic passenger-km (millions) |
|---|---|---|---|---|
| 1 | United States | 765,183 | 214,628 | 550,555 (72%) |
| 2 | Russian Federation | 140,269 | 12,713 | 127,556 (91%) |
| 3 | United Kingdom | 115,199 | 110,462 | 4,737 (4%) |
| 4 | Japan | 108,082 | 55,324 | 52,758 (49%) |
| 5 | France | 58,821 | 37,265 | 21,556 (37%) |
| 6 | Australia | 56,644 | 30,568 | 26,076 (46%) |
| 7 | Germany | 48,965 | 44,083 | 4,882 (10%) |
| 8 | Canada | 41,253 | 24,160 | 17,093 (41%) |
| 9 | China | 40,605 | 11,161 | 29,443 (71%) |
| 10 | Singapore | 37,045 | 37,045 | — (0%) |
| 11 | Netherlands | 32,696 | 32,647 | 49 (1%) |
| 12 | Italy | 28,603 | 21,708 | 6,895 (24%) |
| 13 | Korea | 28,004 | 22,935 | 5,069 (18%) |
| 14 | Brazil | 27,828 | 15,144 | 12,684 (46%) |
| 15 | Spain | 27,480 | 18,198 | 9,282 (33%) |
| 16 | Thailand | 20,427 | 18,616 | 1,811 (10%) |
| 17 | Mexico | 19,553 | 9,119 | 10,434 (50%) |
| 18 | Scandinavia | 18,894 | 13,036 | 5,858 (32%) |
| 19 | Indonesia | 18,737 | 12,076 | 6,661 (37%) |
| 20 | Saudi Arabia | 17,563 | 12,003 | 5,560 (33%) |

*Source*: ICAO (1994)
*Note*: (46%) – Domestic as percentage of total traffic

119

the top 10. The hierarchy for freight performance clearly reflects levels of industrial activity, including recent growth in South Korea.

While Developing Countries have not made a great impact on global air transport overall they feature more prominently in the ranking for domestic services because many such countries are very large, distances are great, centres of economic activity widely dispersed (Brazil, China), terrain is difficult or sea areas intervene (Indonesia) or surface transport is poorly developed (Zaire).

This significance of air transport for domestic services in many Developing Countries is further illustrated by ranking according to number of operational airfields (Table 4.3). Not surprisingly the United States has by far the largest number of public and private airports but possibly of greater interest there are 17 Developing Countries in the top 25. It is significant that many of these countries feature on the list by having large numbers of private rather than public airports – these are clearly the airfields, possibly a more appropriate term than airport, operated by mining, industrial, forestry and agricultural enterprises and other private businesses, presumably in response to the inadequacy of surface transport and public transport provision for their efficient operational needs. In many such cases 'airport' suggests a level of sophistication that is lacking and little more than rudimentary landing strips are provided.

Countries such as Brazil, Mexico and Indonesia appear in the top-20 passenger ranking and they are joined in the airport ranking by a number of Developing Countries characterised by combinations of difficult terrain, long distances and regions in isolation. In Latin America isolation of regions by forest and mountains makes air transport invaluable in Venezuela, Colombia, Ecuador, Peru and Bolivia, while the linear configuration and great distances in Argentina and Chile (ranked 23rd) also favour air transport. In Zaire (22nd) the vast size of the country, the peripherality of main economic regions, the forest environment and the wholly inadequate surface links, and in the Philippines the need to serve the large number of islands which make up the national territory, place a premium on air services. For countries such as Guatemala, Paraguay and Zimbabwe, the placing in the hierarchy is wholly determined by the large number of private airfields, mainly associated with agricultural activity.

*Table 4.3*   Country rank by number of airports, 1989

| | | | Public | | | Private |
|---|---|---|---|---|---|---|
| | | Total | Land | Water | Heliports | Private |
| 1 | United States | 17,167 | 5,268 | 207 | 109 | 11,583 |
| 2 | Brazil | 2,269 | 752 | 0 | 203 | 1,314 |
| 3 | Mexico | 2,042 | 483 | 1 | 112 | 1,446 |
| 4 | Canada | 1,175 | 512 | 78 | 9 | 576 |
| 5 | Bolivia | 1,146 | 590 | 0 | 1 | 555 |
| 6 | Paraguay | 1,001 | 29 | 0 | 0 | 972 |
| 7 | France | 709 | 417 | 1 | 1 | 290 |
| 8 | Colombia | 599 | 201 | 0 | 5 | 393 |
| 9 | Indonesia | 521 | 147 | 0 | 0 | 374 |
| 10 | Zimbabwe | 479 | 33 | 0 | 0 | 446 |
| 11 | Australia | 436 | 430 | 0 | 6 | ? |
| 12 | Papua New Guinea | 435 | 402 | 0 | 0 | 33 |
| 13 | Argentina | 433 | 349 | 0 | 20 | 64 |
| 14 | Kenya | 383 | 152 | 0 | 0 | 231 |
| 15 | Guatemala | 353 | 74 | 0 | 1 | 278 |
| 16 | Venezuela | 333 | 82 | 0 | 16 | 235 |
| 17 | Ecuador | 326 | 49 | 0 | 80 | 197 |
| 18 | West Germany | 316 | 108 | 0 | 2 | 206 |
| 19 | Peru | 297 | 162 | 0 | 1 | 134 |
| 20 | Philippines | 295 | 226 | 0 | 69 | ? |
| 21 | South Africa | 278 | 156 | 14 | 0 | 108 |
| 22 | Zaire | 261 | 100 | 0 | 1 | 160 |
| 23 | Chile | 259 | 129 | 0 | 3 | 127 |
| 24 | India | 225 | 179 | 0 | 0 | 46 |
| 25 | Sweden | 204 | 189 | 0 | 8 | 7 |

*Source*: Based on ICAO (1990)

# THE PIONEER ROLE OF AIR TRANSPORT

The conditions, level of income apart, in north-west Australia in the 1920s were essentially those of many developing regions world-wide at a later date. The distances were great, the terrain difficult, most settlement of significance was peripheral with large areas of low productivity, road and rail transport would have been difficult and costly to provide and the climate, especially flash flooding, inimical to surface transport which was seasonal in character (Holsman and Crawford, 1974).

The first air service was provided along the coast between Geraldton and Derby in 1921 and a 12-day land journey was reduced to 2.5 days by air. This service was government subsidised but even during the depression years of the 1920s and 1930s the air traffic increased as the advantages of the air link were recognised for mail, perishable produce and for the way it reduced general isolation. Aircraft were also used for air surveys, a flying-doctor service was initiated in 1935 and demand for minerals during and then after the war resulted in an elaboration of air services to link new expanding centres of economic activity. The DC3s, which had provided the vital basis for expansion in the 1940s and 1950s, were replaced by higher-capacity aircraft and there emerged in place of the original 'milk run' a more complex pattern of main and feeder routes and services to mines and agricultural stations. The applicability of this model of air transport development to many Developing Countries will be only too apparent.

In 1952, Liberia had only six settlements served by air – five on the coast (Robertsfield, Buchanan, River Cess, Greenville and Harper) but only one in the interior (Tchien). In the 1950s there was a 50 per cent increase in aircraft movements and Puta and Kolahun (not connected by road to the rest of the country) were given air links in 1956 (Stanley, 1965). With the construction of road links to Tchien and Kolahun in the early 1960s the air services to those towns were withdrawn but Nimba and Bomi, two new iron ore mining centres, were added. There was considerable growth of air traffic at Greenville, where the port was being improved, road construction was taking place and where plantations and forest exploration were being initiated.

The development of the Nimba iron ore mine demonstrated the importance of air services. Located in the remote northern border area of Liberia, this was the area of Graham Greene's *Journey Without Maps*, no contour maps existed and there were no direct road links with the coast. An initial air photo survey established the general line for a proposed railway to the coast and a second, more accurate survey with ground control resulted in the determination of the longitudinal profile, grade line and mass earthworks calculations for the line. The area was largely forest covered, with rainfall of over 2,500 mm a year, and helicopters (flying five hours a day) were used to establish the base camps from which the construction work for first a road and then the railway itself could proceed (see Chapter 3). After 1963, the completed railway provided a passenger and freight link between Nimba and Buchanan. As Stanley (1965) observed, air transport in Liberia was a substitute for lack of surface transport.

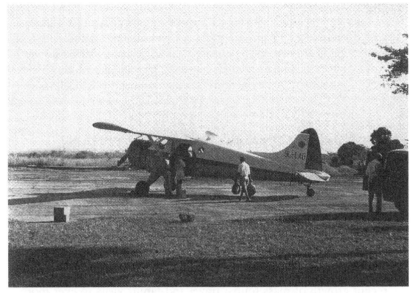

*Plate 4.1* In Sierra Leone, diamond mining operations depend on short take-off Beavers for communication between operating units and for the movement of diamonds, other goods and personnel.

Much the same was true of Sierra Leone, where the first air services were provided by diamond mining companies and in the 1950s short take-off De Haviland Beavers were doing as many as 1,200 flights a year from the capital, Freetown, to up-country mining centres such as Kenema and Yengema. In 1958 Sierra Leone Airways, a subsidiary of the West African Airways Corporation, started domestic services to six principal locations and with landing strip improvements at Kenema and Yengema in 1961, 15-seater Heron aircraft were introduced. The Port Loko to Magbur-aka service ceased soon after but Bonthe and Gbanbatok (close to newly opened bauxite and rutile deposits) were added. Sierra Leone has been described as typifying the pioneer phase of air transport (Reichman, 1965; Williams and Hayward, 1973) with little hope that these air services could survive the improvement of road transport. The scheduled services have disappeared but Beavers are still used and provide the most effective way of moving personnel, money to pay miners, diamonds for export and for getting mail and emergency supplies to the diamond dredges, which are isolated and widely dispersed.

The large number of private airfields in Developing Countries already noted are clearly of the kind just described and provide similar pioneer services to a variety of economic enterprises not easily linked by surface modes. In Chile, passenger and perishable freight movement over one-third of the country is possible only by air.

## AIR TRANSPORT AND DEVELOPMENT

There is a demonstrable positive relationship between the overall use of air transport and levels of income, urbanisation, industrialisation, education and leisure time yet the relationship is multifaceted, complex and far from being a one-way process. The speed of air travel can improve personal welfare by, for example, providing access to medical services, perishable produce and mail and employment opportunities may be widened by speed of travel and mobility. The recreational benefits are clearly demonstrated by the recent growth of mass tourist travel and family visits to ever more distant places and air charters have become important for moving pilgrims to places such as Lourdes and Mecca – large numbers of West African Muslims make the journey to Mecca every year by air. The benefits to society are even more wide-ranging (Wells, 1984) and are summarised in tabular form in Table 4.4. From the point of view of the Developing Countries certain of the benefits of air transport may be highlighted.

Given the geographical characteristics of many Developing Countries even national unity itself is not easily achieved (Zaire, Indonesia) and air transport may provide the only effective means of integrating regions administratively and economically. Many Developing Countries are subject to natural hazards of storm, flood, drought, famine and earthquake and reaching the affected population is often difficult or even impossible by surface transport. The importance of air-lifting of emergency aid has been well illustrated during drought and famine periods in the Sahel and Sudan, after all too frequent floods in Bangladesh and following earthquakes in various places in South America and Asia.

From the point of view of the on-going process of overall development it is the possible role of air transport in stimulating economic activity which is of significance albeit a word of caution

*Table 4.4*   Benefits of air transport

| WELFARE | INDUSTRY |
|---|---|
| Distressed area emergency | Emergency shipments |
| Supplies to isolated communities | Control of decentralised industries |
| Tourism | Access to new markets |
| Foreign product stimulus | Delivery reliability |
| Higher incomes – GNP | Lower inventory costs |
| Employment in airport region | Lower freight and insurance costs |
| | Labour interchange |
| | Cost/time savings for executives |
| | Aerospace industry |
| | Airline local purchases |
| | |
| GOVERNMENT | SECURITY |
| Rapid crisis response | Defence readiness |
| Ease of contact/unification/control | Unity of law enforcement |
| Official communications | Trained personnel (e.g. pilots) |
| (e.g. diplomatic bag) | Rapid domestic disorder response |
| Greater stability | |
| Contribution to balance of | |
| payments | |
| | |
| INTERNATIONAL | |
| Tourism | |
| Diplomatic/official links | |
| Fast overseas freight/sales | |

*Source*: Modified from Wells (1984)

has to be introduced. There is a school of thought which would maintain that closer integration into the world economy, easier access for multi-national corporations and more tourists are all ways in which countries are held in a state of dependency and underdevelopment, the benefits often being illusory. This said, air transport has been the means whereby, for better or for worse, many Developing Countries have been able to expand the range of their economic activities.

Perhaps for Developing Countries some of the general benefits of air transport such as the value of time to the businessman, pleasure flying and entertainment value (air shows, sight-seeing) are of less immediate significance. Of greater concern will be the specific benefits of general aviation (Wells, 1984) in the area of air surveys amd mapping, wild life surveys and patrols, forestry management, fish spotting and stocking, power and pipeline control, crop spraying (fertilisers, insecticides) and disease control spraying (malaria, river blindness).

A second area of interest for the Developing Country will be the generation of business activity. The general aviation services will themselves stimulate employment and spin-off effects in the vicinity of airfields and as already noted may also be the means whereby economic activities (mining, forestry, agricultural) and infrastructure (roads, railways, pipelines, power transmission lines) are developed in remoter areas. A main function of transport is to move goods from areas where they have low utility to areas where their utility is higher – the market for many of the products of Developing Countries is overseas in the more advanced economies and air transport may provide the only possible access to such distant markets. In this way the Developing Country is able to engage in a range of production which its own demand and markets would not justify. Examples will be given below when air freight is considered.

Airlines have often been to the forefront in developing support for travel industries (Wells, 1984) in the form of hotels, ground transport and linking air services. In many West African countries the first large hotels were airline-owned and provided the starting point for the later growth of wider tourist industries; in countries such as Kenya, The Gambia, Thailand, Indonesia and many islands of the Caribbean, tourism based on air travel has come to dominate the economy and allows them to compete long-range with centres traditionally much closer to sources of tourists. Tourism provides 50 per cent of the GDP of the Seychelles, accounts for 20 per cent of all employment and 70 per cent of foreign exchange but only developed after 1971 with the opening of the international airport. Despite its remote location, air transport enables it to cater for a high income tourism based on exclusivity. The growth of Bangkok as a major world airport is largely based on the number of air tourists attracted to Thailand from Japan, Europe, Australasia and North America.

Airports themselves are significant 'growth poles' and around them the services for airlines (fuel, servicing, catering, administration) and the services for passengers (hotels, shops, transport) and freight (storage, customs, agents/forwarders, transport) generate considerable employment opportunities. As a general rule of thumb it has been estimated that every million passengers generate 1,000 airport and 200 off-airport jobs (Hoare, 1974). In addition, for every 1,000 employed in aviation-related jobs there could be as many as another 600 in employment related to the general population and urban growth in the vicinity of the airport. It will be clear that airports attract economic activities, add greatly to the value of their locality and have a considerable multiplier effect and positive environmental impact.

However, there is also a related and significant negative environmental impact associated with noise, congestion and the fact that an airport precludes certain other activities in its vicinity and takes up large amounts of land both of itself and for related industries, housing and access routes. Also, large-scale airport investment may have a negative economic impact to the extent that it utilises capital that could have produced a better return used in some other way. The aviation industry is very capital intensive and has to be developed with considerable thought for the opportunity costs.

## AIR TRANSPORT – THE INTERNATIONAL MODE

Of all the modes of transport, air is by far the most 'international' in its organisation and functioning, depending, as it does, on a high level of reciprocity. At its most basic this relates to the five freedoms of the air (Vance, 1986) – the right of transit through air space, the right to make technical stops (fuelling, emergency), the right to discharge passengers in the country involved, the right to pick up and return to country of origin and the right to pick up originating passengers for carriage to a third country. All of this is ordered by a vast amount of national and international regulation. Air transport is by far the most regulated of the modes both in an economic and technical sense (Doganis, 1985).

In an economic sense there is a high level of regulation of fares, freight rates, frequency of services, capacities and pick-up points. Technically, there is regulation of safety requirements, airworthiness, maintenance, operating conditions, crewing, flight procedures and services (meteorological, air traffic control). In 1944 a conference in Chicago led to the creation of the International Civil Aviation Organisation (ICAO) which became a specialised agency of the United Nations concerned with the development and planning of international air transport according to prescribed principles (Wells, 1984). It is essentially a governmental organisation.

*Plate 4.2*   In desert Mauritania, charter air taxis provide essential business links, Air Mauritanie provides domestic services and the multi-state Air Afrique offers international long-haul connections.

In 1945, the International Air Transport Association (IATA) was created to provide the means of cooperation between air transport operators and is open to all scheduled carriers.

Many of the operational arrangements are based on bilateral agreements between the contracting states of ICAO – this applies to reciprocity, fair exchange of rights, detailed routeing and stopping points and the frequency and capacity allocation to airlines. While they have little option, Developing Countries work within the ICAO framework and in general they favour a system which allows them to participate and share the economic benefits from which they might otherwise be excluded were market forces to be wholly dominant (Naveau, 1989). They see in the ICAO a forum where they have some political influence by force of numbers and they have pressed for less discriminatory access to markets, a more even distribution of resources and aid to allow them to reach the point at which they can compete on equal terms. In general, Developing Countries favour a multilateral tariff coordination and unilateral action such as deregulation by the United States was opposed. They have also supported the idea that equal opportunity should mean equal benefit – this means that even where the provision of services between the two parties is not equal, the financial yield from such services should be divided equally.

Yet the regulatory system has disadvantages in that access to markets is restricted and particularly the level of output of each airline is not wholly under its own control and it has little freedom in pricing. Also, while in theory Developing Countries have an equal chance to compete, the extent to which they do so will depend on the means available to them – and, as in other spheres, these are far from equal. However, it has been suggested that a Developing Country might trade off some traffic rights in order to gain, for example, training facilities for its personnel from a more developed partner (Naveau, 1989).

Many Developing Countries have in fact benefited from ICAO funding for aviation development and this has taken a variety of forms. Often it is in the nature of 'missions' by the ICAO Technical Assistance Bureau. In 1990 over 450 'experts' were employed, many of them on assignment to UNDP projects, and in 1989 nearly 1,500 students from 110 Developing Countries received fellowships for study/training programmes abroad. ICAO has been actively concerned with establishing training institutions in different regions (e.g. the East African School of Aviation, the Nigerian College of Aviation Technology) and a Civil Aviation Purchasing Service assists countries in acquiring sophisticated equipment. In Chile ICAO has been helping with equipment modernisation and in Ecuador assisting with upgrading of aviation infrastructure (e.g. fire-fighting services for smaller airfields in the Amazon region). In Tanzania a Civil Aviation Directorate has been created and in the Philippines a rescue services study has been undertaken. These are just a few of the many projects with which ICAO has been involved; aviation is such a high technology industry and so demanding in skill availability and training, areas in which so many Developing Countries are seriously deficient.

## AIR TRANSPORT SERVICES

The aviation industry is very demanding in its capital investment requirements, fuel costs are high and for many Developing Countries this is an import cost, there is lack of skilled personnel and the competition is from large airlines with vast resources.

Many Developing Countries have ageing fleets, sparse route networks and narrow market penetration and there is the apparent paradox that, while global aviation growth levels have been high, the profit margins have everywhere been marginal and in many sectors in deficit and many airlines, even those of advanced economies, are heavily in debt (Doganis, 1985). Hardly, it might be thought, the type of enterprise from which Developing Countries can gain benefit.

Yet, as already noted, there are large parts of the developing world where it is simply a case of air transport or no transport and the governments have supported an aviation industry usually by direct public ownership of airlines. It has been suggested that in many Developing Countries there is a complementarity of state and airline and a 'nationalist' urge to have a 'flag' airline (Naveau, 1984). There are, however, a number of ways and levels at which the Developing Country can become involved in aviation.

At the simplest, they may leave aviation to local, private initiatives and as already noted there are many business enterprises which have their own air strips and aircraft. The aircraft are usually of the small type (Table 4.1) such as Beavers, Cessnas or Pipers with fixed undercarriages for simplicity of operations and maintenance and an ability to take off and land in short distances and on partially prepared surfaces.

Apart from such 'own-account' operators there are also a variety of air taxi and charter companies using smaller aircraft for survey, agricultural purposes and general passenger and freight movement and together the contribution of such operators is often considerable. In 1985, out a total of 7,287 aircraft registered in Brazil, some 7,110 or 97.6 per cent were under 9,000 kg maximum take-off weight. It would certainly appear that small aircraft use has become very important in Latin America where distance, terrain and large scale agricultural and forestry enterprises combined with North American influence to encourage aviation. In Papua New Guinea, the Philippines and island areas of the Caribbean and Pacific small aircraft use is also favoured.

The next level of involvement would be the provision of domestic services, very often on a hub-and-spoke basis from the capital city to regional centres and as a government service this can be justified as essential infrastructure where other transport is inadequate. Sierra Leone Airways was set up jointly by the government (57 per cent stake) and the then British Caledonian group specifically to provide domestic services and Malayan Airlines was initiated in 1967 to provide similar services. Even after the creation in 1972 of the Malaysian Airline System there was still seen to be a major role in integrating peninsula Malaysia with East Malaysia and the domestic services received special attention, with seven mainland airports linked to four in Sabah and three in Sarawak. Leeward Islands Air Transport (LIAT) started non-scheduled, inter-island links in 1956 and gradually extended its operations, initially in association with the then British West Indies Airline, later with the Court Line holiday and hotel group, whose main objective was getting tourists to its hotels and therefore somewhat at variance with government objectives, and latterly with CARICOM (Caribbean Community) support and Canadian aid and management.

In the late 1980s the Vanuatu government merged two existing private domestic carriers with government support and 26 outer islands, from a total of 80 inhabited islands, are now linked to the international airports at Santo and Port Vila (Barlow, 1990). Shipping services were infrequent and also unreliable and air transport the only feasible way of providing necessary links although the cost is high because many of the islands have only rough grass landing strips and maintenance and spares costs are high, there is high dependence on expatriate staff and few economies of scale with the small aircraft used. This would be a fair description of many of the smaller aviation operations in Developing Countries and the decision to initiate such services may have to be made on social and political grounds rather than pure economics although in the longer run there may be anticipation of economic gain.

Services of this intermediate type in fact use a range of aircraft. Where the demand is low and conditions difficult, rugged, box-like aircraft of the Pilatus Britten-Norman Islander (over 1,000 delivered) and Trilander design have become popular. These are propeller, fixed undercarriage aircraft of 10–14 seats and with a short take-off capability. The De Haviland Twin Otter (Table 4.1) comes in the same category.

Moving up the hierarchy there are still some DC3s in operation and their 'indestructible' nature (Field, 1989), cheapness of operation and ease of flying has never been surpassed. The Hawker Siddeley 748 and the Fokker 27 are the closest more recent equivalents to the DC3 and while rather larger, the British Aerospace 146 has rugged design, low maintenance costs and its four engines give relatively short take-off capability and safety over sea and difficult terrain. These aircraft have a high level of 'systems redundancy which allows minor unserviceabilities to be ignored for a time' (Field, 1989) – this has clear advantages where spares are not readily available and maintenance not as regular or as thorough as might be ideal.

It might be suggested that provision of regular links within countries and between near neighbours and services which are appropriate to demand is as much as many Developing Countries can feasibly sustain in relation to their resources, both financial and technical. Yet many of them have tried to move on to higher levels of aviation provision.

## THE NATIONAL AIRLINE

In 1958, just one year after obtaining full independence, Ghana withdrew from the West African Airways Corporation and established its own national airline, Ghana Airways, to provide domestic

*Plate 4.3*  A Pilatus Britten-Norman Islander provides invaluable and otherwise extremely difficult links with inaccessible communities in Papua New Guinea (Photo: Pilatus Britten-Norman).

and international long-haul services. Nigeria followed Ghana in withdrawing from WAAC and in association with the then British Overseas Airways Corporation (BOAC) and the Elder Dempster shipping line created an independent Nigerian airline. The interests of the two partners were taken over in 1959 and Nigerian Airways came into being as a government enterprise. A not dissimilar pattern of development is common in many Developing Countries with airlines often coming into existence as joint ventures with an established carrier, then becoming independent but all too frequently at some later date having to return to a bigger airline for management and technical support.

There is obviously a desire to 'fly the flag' but it is not easy to assess the extent to which non-economic factors have or indeed should influence governments in their decision making. Prestige is often said to be a motivating factor and can be rationalised in many African countries as an attempt by arbitrarily defined states to create symbols of national, technical achievement and even unity. The Cameroon Tribune (quoted *West Africa*, 5/11/1990) argued against the sale of the national airline's Boeing 747 which was 'a symbol of national pride in the eyes of Cameroonians, at the same level as Yannick Noah, Manu Dibango and the Cameroon football team.'

A second factor may be political in a rather different sense. Ghana Airways, in the early 1960s, expanded its fleet and services to strengthen links with countries of like political view – Congo (Brazzaville), Mali, Guinea, Aden and the Soviet Union. The airline was an instrument for the furthering of President Nkrumah's goals of pan-Africanism and strengthening the bonds between Non-aligned Countries but many of the routes created for these ends operated with very low load factors and on his overthrow, in 1966, the airline's fleet was immediately cut from 18 to six. The location in Addis Ababa of the headquarters of both the United Nations Economic Commission for Africa and the Organisation of African Unity was certainly a factor in the growth and even relative success of Ethiopian Airlines, the emergence of that city as a major hub in the African air network and a stimulus to other African countries to provide long-haul links to this critical political centre.

There are also strategic considerations which have some validity. Newly independent countries in various ways try to reduce the extent to which they are beholden to others. There is a feeling, probably with some justification, that in certain political situations the Developing Country could not depend on the continued provision of air services by operators based in the developed world. There are also the emergencies which need rapid response, such as Ghana's need to bring its nationals out of Nigeria in a hurry in 1983.

There is a sense in which this becomes mixed with economic motives and the desire for import substitution, whether of goods or services. Paradoxically, such import substitution often results in more rather than less dependence, based as it often is on loan finance and considerable imported inputs of equipment, skills and services; this is inevitable for a capital-intensive, high-technology industry such as aviation, especially where that is closely regulated at the international level. The economic motive for becoming involved in long-haul aviation was primarily that of saving foreign exchange and possibly of earning some as well. While profit making on such routes with high load factors is not impossible there is intense competition for traffic and any profits will soon be dissipated if loss-making domestic routes have to be subsidised (Hofton, 1989).

Employment generation is always to be encouraged in Developing Countries and in a number of smaller countries the national airline is often a significant employer (Table 4.5). The staff costs will depend on the extent to which skills (especially flight-deck crew and maintenance) have to be imported and this is high with respect to many African countries but low for those of Latin America and India. The proportion of imported maintenance staff may depend on arrangements with larger airlines and the demand for cabin crew will be less on domestic flights (e.g. Indian Airlines in contrast with long-haul Air India) especially where distances are relatively short. There has certainly been the suggestion with respect to some African airlines that they have been over-staffed and perhaps significantly Nigerian Airways in 1987 had the highest proportion of its staff in the vague 'handling' and 'undefined' categories.

The really successful long-haul airlines – Cathay Pacific (Hong Kong), Singapore Airlines, Thai Airways, Korean Airlines, China Airlines (Taiwan) and the recently created Eva Air (Taiwan) – are all based in strong export economies with no problems in generating foreign exchange, a high level of self-sufficiency in staffing, relatively low operating costs and with network hub locations and/ or strong tourist demand (Hofton, 1989). A high proportion of the routes they serve have very high load factors and in the 1970s and

Table 4.5    Selected airline employment, 1987

| | Pilots/ co-pilot | Other cockpit staff | Cabin staff | Maintenance staff | Ticketing, sales, promotion | Airport, handling | Other | Total |
|---|---|---|---|---|---|---|---|---|
| Air Botswana | 13 | — | 15 | 22 | 43 | 67 | 51 | 211 |
| Air Malawi | 14 | — | 21 | 134 | 94 | 136 | 423 | 822 |
| Air Gabon | 38 | 7 | 97 | 226 | 275 | 527 | 360 | 1,530 |
| Air Peru | 97 | 32 | 216 | 374 | 388 | 162 | 422 | 1,691 |
| Avianca (Colombia) | 224 | 114 | 468 | 1,310 | 363 | 338 | 2,197 | 5,014 |
| Air Afrique | 90 | 55 | 453 | 640 | 792 | 2,118 | 1,530 | 5,678 |
| Nigerian Airways | 220 | 38 | 498 | 678 | 450 | 2,626 | 2,122 | 6,632 |
| Garuda (Indonesia) | 582 | 82 | 1,638 | 1,810 | 799 | 1,567 | 2,333 | 8,811 |
| Aerolineas Argentinas | 546 | 175 | 1,393 | 1,703 | 1,153 | — | 5,314 | 10,283 |
| Mexicana | 1,006 | 140 | 1,775 | 3,010 | 2,108 | 3,805 | 2,208 | 14,052 |
| Air India | 271 | 129 | 1,923 | 3,497 | 3,050 | 1,123 | 7,397 | 17,390 |
| Indian Airlines | 451 | 49 | 956 | 7,594 | 1,637 | 3,644 | 5,846 | 20,177 |

*Source:* IATA, 1987

1980s, Singapore, Thai and Korean airlines had the highest growth rates of any. These are also the airlines of what might be termed the Newly Industrialised Countries and tend to confirm that successful airlines are usually associated with successful economies – they are not typical of Developing Countries in general which have few, if any, of the advantages just noted.

## AIRLINE COSTS

Even for the biggest and the best, life in the airline business is not an easy way to profit. Rapid growth in demand was matched by technological change especially the introduction of larger aircraft. Airlines were faced with fuel price rises (1974, 1978/9), recession, deregulation, growing competition from charter operators and escalating capital costs (Doganis, 1985). Thus, whereas a 1950s replacement aircraft might have cost $2 million, in 1974 a Boeing 727 cost $8–9 million and in 1984 an A310 Airbus was $45 million. More stringent noise regulations now in operation mean that aircraft such as the DC8 and 707 which were the basis of many Developing Country fleets have had to be replaced. With high interest rates this kind of investment is beyond the means of most Developing Countries and most airlines are heavily in debt – a fact that must be taken into account. Broadly, airline costs can be divided into direct operational and indirect costs (Table 4.6).

The crew costs include salaries, expenses and stop-over costs and while the salaries may be paid in local currency the other elements may well be in foreign exchange. Fuel costs will vary greatly with aircraft type, sector distances, cruise altitude and weather conditions and much of this will be in foreign currency – in non-oil-producing Developing Countries fuel will be an import cost even if the airline pays in local currency. Most of the airport landing fees and en-route charges (usually related to weight and distance covered over airspace of country) will be foreign exchange costs. The depreciation, amortisation and debt charges will be foreign currency and elements of each of the indirect costs may also not be local charges. It is important to note that the fuel costs have been increasing rapidly and overall it seems that the airlines of most Developing Countries, far from being generators of foreign exchange, have been a continuing drain on scarce resources.

It is thought that economies of scale within airlines are weak so there is not necessarily any gain in trying to become large and most

*Table 4.6*  Airline operating costs (percentages of total)

|                              | 1972 | 1982 |
|------------------------------|------|------|
| *Direct*                     |      |      |
| Flight operations, of which: | 29.8 | 41.7 |
|   crew             | 10.1 | 7.3  |
|   fuel/oil         | 11.0 | 27.2 |
|   airport/en route | 3.7  | 4.7  |
|   insurance/rentals| 5.0  | 2.5  |
| Maintenance                  | 13.8 | 9.8  |
| Depreciation                 | 10.6 | 6.8  |
| TOTAL                        | 54.2 | 58.4 |
| *Indirect*                   |      |      |
| Station/ground               | 14.0 | 10.8 |
| Passenger services           | 10.1 | 9.1  |
| Ticketing/promotion          | 15.1 | 15.5 |
| General administration       | 6.6  | 6.1  |
| TOTAL                        | 45.8 | 41.6 |

*Source*: After Doganis (1985)

of the costs tend to increase more or less in line with scale. There is reason for believing that some of the airlines have been over-administered, a problem not uncommon in government enterprises and that these costs have not been held properly in check. There are, however, strong economies between aircraft and the number of aircraft types in the fleet should be kept to the minimum required to satisfy the different types of demand in terms of capacity, distance and route and airport characteristics. This produces economies in maintenance, reduced inventory of spares and ease of transfer of parts between aircraft in emergencies.

Overall, the cost will be related to a number of operating parameters. Much depends on the aircraft utilisation rate and this is related to the length of haul – over short distances the 'block' speed is low with the aircraft spending less time at economical cruising speeds, aircraft and crew utilisation is low and frequent take-off and landing, especially on poor runways, increases fuel costs and also maintenance. Thus, the operation of short-haul, frequent stop domestic routes which characterise so many Developing Countries is rarely a low-cost option even if a higher proportion of the costs are in local currency.

The economics of airline operation depends very much on the

load factors and many of the costs – fuel being the obvious exception – do not vary significantly with load. Globally in the late 1980s passenger load factors averaged about 68 per cent for domestic and international flights but there are considerable variations between routes and many of those served by the airlines of Developing Countries tend to be at the lower end of the scale. The lowest load factors for passengers tend to be the routes within Developing Countries (Africa, South America) or between developing regions (e.g. Central America – South America). It tends to be the same routes which have the lowest weight utilisation rates and this must reflect the well-known fact that communication and trade between Developing Countries is much less developed than that between the developing and advanced economies. There is obviously excess capacity on many South–South routes and these are just the ones which the larger airlines do not wish to service and which are left to the Developing Countries themselves.

Nigerian Airways, albeit an extreme case, serves to illustrate some of the problems. In 1982 it could be claimed that it was the fastest-growing IATA member (*Lloyd's List*, 4/3/1982), probably a reflection of increased government funding following the oil price increases of the mid-1970s and the resultant big increases in passengers and freight which could not be accommodated at heavily congested seaports. Yet the airline failed to make any money, at least in part because the government's attempt to provide a social service on domestic routes forced the airline to charge fares which were supposedly less than for taxis over equivalent distances. The airline was also forced to purchase aircraft types determined by government funding arrangements rather than those most suited to their operations.

On appointment to a management contract in 1979, KLM found that no accounts had ever been published and there was a huge backlog of unsettled debts for passenger and freight movement on behalf of government departments. KLM improved services but also committed the airline to nine B-727s for domestic services and two B-747s and four Airbus A310s for international routes. The KLM contract was terminated in 1981. By the late 1980s services had deteriorated, cancellations were frequent and fares had been sharply increased. In 1987 out of a fleet of 18 aircraft only eight were reported operational, there was no functioning repair hanger, while lack of foreign exchange and spare parts and competition from some new private carriers on domestic routes combined to reduce traffic. In 1984 there was a massive shedding of staff followed almost immediately by even more appointed. Moreover, all too frequently Nigerian aircraft are arrested at foreign airports for non-payment of debts. Against this background the airline was trying to service 16 domestic, 12 West African, four other African, three European and one North American routes. In 1988 a Presidential Task Force was created to sort out the airline's problems but achieved little.

It is not known what became of a suggestion that the Nigerian Air Force should start unscheduled flights at cheap rates for students, military personnel and general public but the three private companies have also run into difficulties in consequence of rising costs, government directives over routeing and ageing aircraft, which have become known as 'molues of the air' – the molues are the extremely dilapidated Lagos buses (*West Africa*, 3/6/1991). It might jokingly be remarked of Nigerian Airways that one should not expect too much of an airline which for long had as its symbol an elephant with wings although there is no evidence that the recent adoption of an eagle has improved its fortunes. Yet if Nigeria, with the largest population in Africa, its considerable oil revenue and relatively high level of industrial development cannot support an airline, there is little hope for its poorer neighbours.

On purely economic grounds it is impossible to justify the operation of such a national airline and while it too has had financial problems, the multinational Air Afrique, created originally by 12 Francophone West African countries, might provide a better model for government involvement. Not surprisingly, in 1992 plans were announced for the privatisation of Nigerian Airways but as yet there has been little progress in this direction. As state-owned enterprises some of the airlines have been over-protected, over-controlled and under-financed and there are strong arguments for central government divesting itself of this costly burden but possibly creating strong, autonomous aviation administrations to regulate and encourage the private sector. With UNDP and ICAO assistance Uganda is moving in this direction and there has already been good progress towards the rehabilitation of airfields and services (Diallo, 1994). However, there are few Developing Countries in which the domestic private sector would be able to finance airline operations on any scale.

# AIR FREIGHT

In the early days, airlines were heavily dependent on mail contracts to support passenger operations but in recent years the air cargo market has developed as a function of growth in passenger travel. The aircraft fuselage shape leaves a great deal of spare 'belly' capacity under the cabin floor – in a 747 this has a weight capacity of 18 tonnes after passengers and baggage have been accommodated. Because air freight was a by-product of passenger operations and also because its revenue yield was less it was for long viewed as the poor relative of the passenger services and was dominated by what has been termed the 'Rembrandts and racehorses' philosophy – air transport was for the occasional movement of goods with high value in relation to weight, or emergency movements (e.g. spare parts for machinery or equipment repairs where time saving was critical and expense of secondary significance). It was essentially a premium service at high price – a powerful psychological barrier to its expansion (Hayuth, 1983).

A number of factors operated to change this view of air freighting. In 1948 the nine-month blockade of West Berlin and the resultant air-lift demonstrated that air transport could carry large volumes of goods including industrial raw materials. In the late 1950s the advent of freight versions of the DC8 and 707 gave pay loads of 35 tonnes and the later 747 increased this to 120 tonnes. The unit cost of air freight declined dramatically – from 35.9p per capacity tonne mile in a Britannia to 16.5p in a 707 and only 11.1p in a 747. In freight only configurations the drop was even more marked. The greater use of wide-bodied aircraft for passengers added significantly to freight capacity and reduced the need for freight-only services on many routes although the total air freight market has been expanding at about 7.7 per cent a year from the early 1980s.

In contrast with passengers, freight is not self-stowing (Taneja, 1989) and improvements in cargo handling have been a necessary factor facilitating cargo growth. On aircraft with freight configurations, wide side or nose doors and deck rollers make possible the handling of pallets or even full-sized containers while many airlines now use Unit Load Devices (ULDs), containers shaped to fit into different parts of the hold and therefore of variable capacity.

The real selling point for air freight is the time savings that can be effected and air transport is now seen as a part of the total distribution chain and cost. Regular movements by air allow justin-time delivery and reduced cost of stock holding (space, staff, paper work). Air transport does not require heavy packaging and insurance rates are lower than by sea. Thus, while at first sight air freight rates may appear much higher than those for surface modes, the total cost may well be less. Viewed in this way, air transport becomes feasible for a wide range of commodities and many Developing Countries have been able to capitalise on this to expand the range of their exports (Figure 4.3). With frequent scheduled flights on many routes, goods can be put on the market very quickly and this is absolutely critical where goods are perishable and have short shelf lives or where the economic life of the commodity is limited, as with fashions or the Christmas trade.

In particular, air-freighting allows Developing Countries to compete in markets that would not otherwise be open to them. Every week, some 400 tonnes of fruit and vegetables are airfreighted into London from African and Caribbean countries alone and British supermarkets are now stocked with a wide variety of fresh goods from the tropics – pineapples from Côte d'Ivoire and Malaysia, green beans from Kenya, ugli fruit and bread fruit from Jamaica. Mozambique sends shrimps and Côte d'Ivoire lobsters to France and the Seychelles exports red snapper fish. Côte d'Ivoire, Kenya and Colombia send cut flowers to London, often timed to meet specific market windows. In Zimbabwe, intensive farming of fruit and vegetables was in 1995 expected

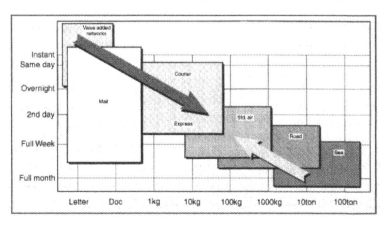

*Figure 4.3* Time and freight transport
*Source:* ICAO

141

to produce £35 million worth of exports and rapid growth of this sector of the economy has only been possible through air freighting of mangetout, sweet corn, baby corn, passion fruit, mangoes, Cape strawberries, raspberries, plums and cut flowers to supermarkets in Europe. From picking in the field, packing, labelling and bar coding to supermarket shelf can be as little as 24 hours and at most is three to four days.

Botswana exports prime beef to Switzerland and the preparation of chilled choice cuts adds to the value of the product, creates employment and produces a commodity that is more easily transported than carcass meat. That said, Chad, Colombia and the Rupununi region of Guyana have all used air transport for frozen carcass export, in Chad's case mainly to the larger coastal markets of Douala and Brazzaville. Farm produce is the fastest-growing sector of Thai air exports with 10,000 tonnes a year of fruit such as longons, lychee and grapes to Hong Kong alone.

Less perishable goods may also be air-freighted to avoid long land hauls and usually to satisfy fairly specific markets. Uganda has used air transport for tea to Manchester and Zaire for its coffee to Mombasa for onward movement by sea, or on occasions, directly from Goma to Britain. In some cases the movements are 'domestic' in character. In Vanuatu, seafood is air-freighted from outlying islands to Port Vila and in Ghana there were occasions when military aircraft were used to ferry tomatoes from the north of the country to a canning factory in the south.

The role of air transport in allowing market penetration for the agricultural products of Developing Countries is clearly invaluable but there has also been growing use for industrial semi-finished and finished goods. Over 70 per cent of India's textile and leather goods exports are by air and this has been associated with a move into higher-quality, more discriminating fashion markets. Motor vehicle engines for cars assembled in India have sometimes been air-freighted in to avoid delays at seaports. India is one of the few Developing Countries in which air freight capacity is used virtually to saturation point and the government has considered the creation of a cargo-only airline. Air India and Indian Airlines have both converted older jets into pure cargo carriers and in 1994 Madras-based Patel On-Board Couriers set up the country's first freight airline. Indian air freight exports were expected to amount to 36,000 tonnes in 1995 while domestic cargo would be as much as 181,000 tonnes.

*Plate 4.4* Loading Unit Load Devices with freshly picked mangetout and runner beans at the Utopia Fresh Produce plant in Zimbabwe (Photo: Barry Deacon, Utopia Fresh Produce).

For many years, Peugeot's car assembly plant at Kaduna in northern Nigeria received all components by air and this only stopped when port handling facilities for containers and onward transport by rail were improved and delivery service became acceptable. In Niger, the remote uranium mining centre at Arlit has frequently used air transport for equipment and processing materials and the small quantities of high value uranium concentrate are air-freighted out.

Air transport has also made possible some interesting, although some would consider exploitive, industrial relationships. In St Lucia, the Inner Secrets garment manufacturer of Hoboken, New Jersey, set up a factory which employs 460, the material being bought in the United States, cut in Hoboken, air-freighted to St Lucia via Trinidad for making up (capitalising on cheap labour availability) and the finished garments then return by air to America for distribution. Apparently, with a correct sense of priorities, shipments were limited during the Trinidad carnival period to make way for more important traffic! Plastic boxes made in Brazil are air-

143

freighted to Taiwan where locally manufactured shoes are put into them for air-freighting to San Francisco. Such global inter-industry movements require fast, frequent and reliable transport which only air transport can provide over the distances involved.

Air transport still has its important role in the emergency movement of freight and there are all too frequent reminders that for Developing Countries this is often for disaster relief, where humanitarian rather than cost considerations are dominant. High-value goods such as precious stones or bullion are able to 'bear the cost' of air transport without problems but by allowing cost savings elsewhere (Doganis, 1985), air transport can be used for a wide range of products especially where time and the need to meet market deadlines are important. A study of air freight potential in Africa (Economic Commission for Africa, 1975) concluded that air transport is an essential development tool and considered as a part of total distribution costs not necessarily more expensive than other modes. Its ability to open up new markets and stimulate productive activity is its main advantage but the strong links with metropolitan countries has meant a neglect of intra-African services. Main impediments to further development of air freight were seen as directional and seasonal imbalances of traffic, inadequacy of terminal facilities and slow customs clearance.

It seems likely that growth of air freight, like that of air transport in general, will be fastest in the Pacific rim region and lowest on routes to and within Africa and Latin America with increasing marginalisation of these regions. There is evidence of concentration into the hands of a small number of large operators or combines which may make it difficult for small and medium sized airlines to compete – this could be to the disadvantage of Developing Country airlines which may have to identify their market niche and concentrate on that area. It will certainly mean investment on their part in handling facilities and information technology.

## AIRPORT FACILITIES

The extent to which a Developing Country is able to become involved in air transport will be related to its preparedness and ability to provide the necessary supporting airport infrastructure. From data presented earlier (Table 4.3) it emerged that many Developing Countries have a large number of small private airfields and these are associated with a high proportion of small aircraft registrations. Many of the airfields are of the most rudimentary kind with short, unsurfaced runways and little in the way of navigational aids and terminal facilities. Such facilities may well be seasonable in availability and closed to traffic in adverse weather conditions. While they may well have a vital role in terms of the enterprise or settlement they serve, and are in this sense appropriate, they clearly have limited commercial potential. There are few countries which do not now have an airport able to accommodate international services, although not necessarily for the largest aircraft, but in some cases these have been inherited from a time when civil aviation was much less demanding. Accra's airport (Ghana) is arguably far too close to the city while Monrovia (Liberia) is served by an airport 80 km away. Lungi Airport (Sierra Leone) is separated from the capital Freetown by an erratic ferry with bus connections at either end.

In 1985 major airlines threatened to cease operations at Lungi, arguing that it had become unsafe. Unreliable electricity power supplies meant that runway lights, navigation aids and communication systems were often not working, the security fence was broken and people had to be persuaded not to use the runway as a road (*West Africa*, 3/7/89). A management restructuring has apparently led to some improvements. In 1989 it was reported that Nigeria's 16 airports were 'crippled by lack of maintenance' and Murtala Mohammed Airport at Lagos had slipped in the ICAO rating from Category 9 to Category 6, the airport being unsafe for the largest jets because of inadequate fire fighting equipment. Again, there were reports of the runways being used as expressways for road vehicles.

## AIRPORT PLANNING

The above examples while probably extreme do indicate some problems faced by poorer countries in providing, maintaining and managing their airports when capital is not available and the technical and managerial skills possibly lacking. Airports from the simplest to the most complex share three common characteristics (Poole, 1990) – they should be appropriate to their role, cost-effective and safe. In order to satisfy these requirements the planning and design of the airport, whether new or an expansion of existing facilities, is of critical importance. The air transport system, represented in Figure 4.4, is a complex of interrelated elements

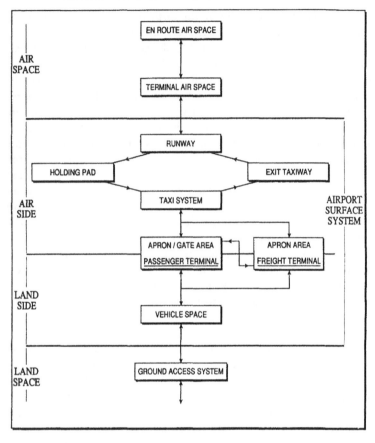

*Figure 4.4*  The air transport system
*Source*: Modified from Yu (1982)

which should ideally be planned as a totality – the capacity of the system will be a function of its weakest element. In the case of some European airports it might be argued that the main constraint on capacity is in the air space or possibly in the ground approaches to the airport while in most Developing Countries, especially smaller ones, the problems are likely to derive from the terminal itself or possibly the runway.

As the starting point for any airport development the need has to be established and quantified (Kaberry, 1989). This will involve consideration of many of the factors already discussed above but particularly the role the airport is expected to play and the present demand and anticipated growth in the markets it is to serve – general passenger, tourist, freight, distribution, industry. The quantified need has then to be translated into facilities – where they should be located and of what type. At this point the airport designer can take over (Kaberry, 1989) to determine the ground plan and the design characteristics of the individual elements – runways, taxiways, terminal facilities and functionally related surroundings.

In infrastructure development there is always the question of precisely how much capacity to provide; if no surplus is built in at the outset later piecemeal additions are likely to be expensive and disruptive of operations and may be difficult or impossible to incorporate. However, excess capacity at the outset will be under-utilised and increase costs. For this reason the location should allow for future expansion and the design should be such that later additions and functional modifications can easily be incorporated. It is essential that other urban functions should not be allowed to encroach on the land that may be needed for runway extension or additions to facilities. For a variety of reasons, but noise in particular, it is highly undesirable to have residential areas, schools and hospitals close to airports and the presence of some types of industry (oil refineries, chemical plant, nuclear power stations) may be hazardous.

## RUNWAYS AND CAPACITY

In many respects the single most important influence on design will be runway length and this is determined by the aircraft types and performance and the effect on that performance of geographical factors such as altitude, temperature, wind direction and surrounding terrain (affecting approach and climb-out angles). Some typical runway requirements are indicated in Table 4.1 The airports of many Developing Countries in east and southern Africa, the Andes and the Himalayas are both at high altitudes and have high temperatures and for any particular aircraft longer runways will be needed.

The runway has to distribute the weight of the aircraft to the sub-surface and much of the discussion relating to railway-track bed design (Chapter 3) is equally applicable to runways. Local aggregates may be unsuitable and imported materials may be necessary. Soil stabilisation and sealing may be needed and drainage to counteract the severity of tropical storms will have to be considered. The actual pavement strength will be determined by the aircraft to be accommodated and can be expressed as the maximum weight of aircraft that can be accepted (e.g. Belo Horizonte, Brazil: 73,480 kg; Dallas, USA: 362,874 kg).

As for any other structure there will be trade off between cost and durability – lower initial capital cost may well mean higher maintenance costs and earlier replacement while any likelihood that larger aircraft may be brought into use should be taken into account at the outset. To runway length and pavement strength must be added the important consideration of runway layout (Figure 4.5). As a means of increasing capacity at existing airports, adding to the number or changing the layout of runways is often impossible for space reasons especially where the airport is adjacent to or surrounded by urban development.

The capacity will also be influenced by the mix of aircraft, whether visual or instrument flight rules apply, the amount of instrumental assistance and, on the ground, the extent to which runways are also used as taxiways to and from terminals. With no separate taxiways, a single runway might be expected to take 10 movements an hour but with parallel taxiways for aircraft landing and taking off the capacity would be 40 movements an hour (Poole, 1990).

The American Federal Aviation Administration has established capacity planning guidelines for different mixes of aircraft (Figure 4.5) and different configurations of runways. Properly used, a single runway can have a considerable capacity – Gatwick, UK, has a capacity of 200,000 aircraft movements a year and a throughput of 20 million passengers. On this basis it has been argued (Poole, 1990) that probably fewer than 100 of the world's 4,000 airports handling scheduled traffic require more than one runway. Many already have more but this is in part because they were built at a time when aircraft performance and instrumental aids were less sophisticated and also because management does not spread traffic evenly over time (e.g. for noise reasons night flying may be restricted). There is also a tendency for longhaul flight arrivals to be bunched for passenger convenience – people prefer to arrive early in the morning rather than late at night! Some airports do not exclude general aviation which might be more suitably accommodated away from main airports.

There are always arguments in favour of spare capacity to meet

| RUNWAY LAYOUT | DESCRIPTION | MIX | PRACTICAL HOURLY CAPACITY | PRACTICAL ANNUAL CAPACITY (AIRCRAFT MOVEMENT) |
|---|---|---|---|---|
| | SINGLE RUNWAY MIXED - MODE | 1 | 53 - 99 | 215,000 |
| | | 2 | 52 - 76 | 195,000 |
| | | 3 | 44 - 54 | 180,000 |
| | | 4 | 42 - 45 | 170,000 |
| Less than 3,500 ft | CLOSE PARALLELS DEPENDENT INSTRUMENT FLIGHT RULES | 1 | 64 - 198 | 385,000 |
| | | 2 | 63 - 152 | 330,000 |
| | | 3 | 55 - 108 | 295,000 |
| | | 4 | 54 - 90 | 280,000 |
| 3,500 to 4,999 ft | APPROACH / DEPARTURE PARALLELS INDEPENDENT INSTRUMENT FLIGHT RULES | 1 | 79 - 198 | 425,000 |
| | | 2 | 79 - 152 | 390,000 |
| | | 3 | 79 - 108 | 355,000 |
| | | 4 | 74 - 90 | 330,000 |
| 5,000 ft or more | ARRIVAL / DEPARTURE INDEPENDENT INSTRUMENT FLIGHT RULES | 1 | 106 - 198 | 430,000 |
| | | 2 | 104 - 152 | 390,000 |
| | | 3 | 88 - 108 | 360,000 |
| | | 4 | 84 - 90 | 340,000 |
| 5,000 ft or more | INDEPENDENT PARALLELS and CLOSE PARALLELS | 1 | 128 - 396 | 770,000 |
| | | 2 | 126 - 304 | 660,000 |
| | | 3 | 110 - 216 | 590,000 |
| | | 4 | 108 - 180 | 560,000 |
| | TWO INTERSECTING, NEAR THRESHOLD | 1 | 71 - 175 | 375,000 |
| | | 2 | 70 - 125 | 310,000 |
| | | 3 | 63 - 83 | 275,000 |
| | | 4 | 60 - 69 | 255,000 |

| MIX | A | B | C | D | |
|---|---|---|---|---|---|
| 1 | 0 | 0 | 10 | 90 | A = 4 engine jet |
| 2 | 0 | 30 | 30 | 40 | B = 2 / 3 engine jet   4 engine piston / turbo |
| 3 | 20 | 40 | 20 | 20 | C = utility twin-prop   executive jet |
| 4 | 60 | 20 | 20 | 0 | D = light twin / single engine |
| | ( % of each type ) | | | | |

*Figure 4.5*  Runway capacity guidelines
*Source*: American Federal Aviation Administration

emergencies and also to facilitate maintenance. Given the potential capacity of even a single runway it is often the terminal facilities that act as a constraint on throughput.

Terminal facilities should be appropriate to the demand, expected changes in that demand, local conditions with respect to formalities for passengers and freight (security, immigration, health, customs), facilities for inter-lining, access to surface transport, parking and climatic conditions. Where rain is frequent and heavy, as in many tropical countries, most operations must be under cover. However, the author has fond memories of sitting out under the palm trees, an open-air extension of the departure lounge, at Banjul, The Gambia, in the pre mass-tourist days.

In the congested skies of Europe and parts of North America there are grounds for arguing that the capacity of the air transport system is determined by the sophistication of the air traffic control system but there are few other areas where this applies. However, from a safety, if not capacity point of view, the adequacy of the air traffic control system is everywhere critical.

The American Federal Aviation Administration assigns airports to 'design groups' based on aircraft wingspan and approach speeds (Figure 4.6), with a broad distinction between 'utility' airports for smaller aircraft with lower approach speeds and 'transport' airports able to accommodate the larger aircraft with higher approach speeds. In effect, this distinction is related to runway geometry, and especially runway length. International airport classifications such as those of the ICAO are based in the first instance on runway length but with refinements based on runway width/ gradient, taxiway widths, glide path slopes, navigational aids (non-instrument, precision approach) and the array of services and safety provision.

Such classifications reflect the increasing differentiation between airports associated with growing technological sophistication and scale of operations. The emergence of hierarchical structures is common in transport but is particularly marked in air transport with its well developed hub-and-spoke patterns of operation. It is often the case that a small number of pivotal, international standard airports provide the 'gateways' (a term used by American airlines to describe their international access points to the domestic system) to regions or nations.

This can be seen in the case of Gabon (Figure 4.7) where the capital, Libreville, and important economic centre, Port Gentil, are

| WING SPAN (FEET) | UTILITY AIRPORT DESIGN GROUP | APPROACH SPEED (KNOTS) | | | | | TRANSPORT AIRPORT DESIGN GROUP |
|---|---|---|---|---|---|---|---|
| | | <91 CAT. A | 91 - 120 CAT. B | 121 - 140 CAT. C | 141 - 165 CAT. D | >166 CAT.E | |
| <49 | I(a) | UTILITY AIRPORTS | | | | | I(a) |
| <49 | I | | | | | | I |
| 49 - 79 | II | | | | | | II |
| 79 - 118 | III | | | TRANSPORT AIRPORT | | | III |
| 118 - 171 | | | | | | | IV |
| 171 - 197 | | | | | | | V |
| 197 - 262 | | | | | | | VI |

(a) Small aircraft only

*Figure 4.6* Airport design classification
*Source:* American Federal Aviation Administration

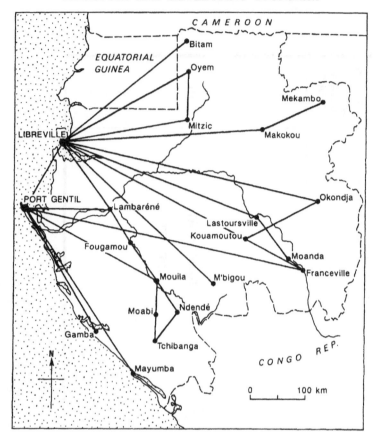

*Figure 4.7* Gabon: hub-and-spoke network
*Source:* Based on *ABC World Airways Guide,* 1991

linked by long-haul international carriers to a variety of other African cities and main centres in North America and Europe. From Libreville and Port Gentil, domestic routes fan out to a number of regional centres and for the latter, with no road links, these provide the only connections to the hinterland. With variations dependent upon size and geographical characteristics many Developing Countries exhibit similar patterns of air transport provision.

## AEROSPACE INDUSTRIES

From the point of view of stimulating development, an important feature of transport is its ability to generate employment in associated industries concerned with the maintenance and possibly manufacture of the equipment or vehicles used. We have seen that the potential is considerable in the case of road transport but less significant with railways. What is the position for the aviation industry?

It will be clear that for other modes there is a range of technological levels at which the Developing Country can become involved and it is not too difficult to do so at a level which is appropriate to its resources and skills, the skills often transferable through traditional forms of apprenticeship (e.g. boat-building skills, road vehicle repairs). The aircraft is in all respects a high-technology product with vast research and development costs and the need for large production runs to ensure a return on capital. It is demanding of a range of skills that are absent in most Developing Countries and only produced at great cost. Even at the level of routine aircraft maintenance it is often the case that technicians have to be trained abroad, foreign technicians have to be imported and maintenance packages arranged with the aircraft manufacturers or larger airlines.

Few Developing Countries are likely to have the conditions which would permit a fully fledged aerospace industry and only a handful have taken steps to develop such an industry. An obvious starting point is the assembly under licence of aircraft or aero engines from imported, knocked-down kits. Colombia, Egypt, India, Indonesia, Pakistan and the Philippines assemble either engines, aircraft or helicopters and Malaysia and Thailand both produce components for export. Argentina produces a two-seat, crop-spraying aircraft with air ambulance and training versions and also military support aircraft and Chile produces a single-engine military trainer. The local military market has often been the vital factor ensuring sales adequate to justify production. In Africa, only Egypt (engine assembly) and South Africa (aircraft assembly) have become involved.

Brazil is the only Developing Country to have created a substantial aerospace industry which has managed to break into foreign markets. Amongst Developing Countries, Brazil is second only to India in the number of trained scientists in research and development and its geographical conditions and considerable domestic civil and military market for aircraft, place it in an exceptional position. Active state involvement started as early as 1941 and in recent years government support has been substantial. Empresa Brasileira de Aeronautica (EMBRAER) came into operation in 1970 and is almost entirely government-funded. By 1983 it had a product line of seven aircraft, 4–5 seater Piper Seneca and Navajo built under licence and five designs developed in Brazil. By the late 1980s over 3,700 aircraft had been built by a work force of nearly 10,000. In addition, three other firms, employing a total of 1,200, produced other aircraft under licence.

The Brazilian military account for 75 per cent of local sales and place large orders 'up-front' to assist with R and D costs. Since 1974 there have been government restrictions on the import of aircraft equivalent to those produced locally and there is a levy on long-distance air travellers which is channelled into loans to support regional airlines using local aircraft. It was felt that this middle-level service, between small aircraft utility flying and big airline operations, was in particular need of development and support (Ramamurti, 1985).

The best-known local aircraft are the various versions of the Bandeirante, the development of the 21-seater Pioneer starting in 1968 and the 30-seater Brasilia in 1983. Originally for military transport, training and aero-medical roles, these twin-propeller planes have a variety of civil roles and over 500 have been produced, nearly 20 per cent being exported to 36 countries, mainly for use in short- and medium-haul 'commuter' services. EMBRAER also produces the Xingu, an executive aircraft, the T–27 military trainer and the Ipanema, a single-engine crop-spraying aircraft. Brazil has succeeded by concentrating on the turbo-propeller market with a good domestic market base and export potential, by emphasising design for specific conditions with frequent upgrading and variants and by avoiding the local manufacture of the higher-technology inputs such as the engines, landing gear and avionics. The import element amounts to between 27 and 41 per cent for the locally designed aircraft and 47 to 71 per cent for those produced under licence (Ramamurti, 1985).

## A CAUTIOUS APPROACH?

It will be clear that while air transport has been a critical element in the internal cohesion, opening up and consequent economic development in many countries it is a mode characterised by technological sophistication, high investment costs and limited potential for local employment in the initial stages and high elements of risk at all stages. It is a mode which has little direct relevance for the bulk of the population except in very indirect ways and can only provide very low density networks and limited access. It is therefore vital for the Developing Country to participate in this transport mode with caution because the impact can all too readily be negative with large amounts of investment needed to provide services which may never be economically viable (Persaud, 1986). It is, of course, open to any government to decide that there are stronger, overriding social or political reasons for developing and supporting such services.

Many Developing Countries have created long-haul air services, usually with state-owned airlines which have found it difficult to compete with the usually much larger, longer-established airlines based in Europe or North America and have had severe problems in maintaining standards where resources are scarce. Where there has been profit on the long-haul routes it has often been used to subsidise loss-making domestic services (Hofton, 1989). It has been argued that the best way for smaller African airlines to survive is to 'join forces, maximise their size, maximise their capability (and) governments must learn not to interfere with commercial management of the airlines' (N'Diaye, 1988).

This said, Air Afrique, a multinational enterprise serving former French African territories, has experienced severe financial problems and in the late 1980s was forced to restructure its operations; 2,000 staff were made redundant and it generated hostility from competing airlines by restricting their flights and passengers per flight. Given that air traffic agreements are between countries and not airlines, Air Afrique's actions were of doubtful legality and brought reprisals – for a time Italy stopped Air Afrique using Italian airports. Where the airlines are state-owned there is the danger that government departments use them as if they were free services – in 1987 Air Afrique was owed $40 million in unpaid fares and this has also been a main problem for the national airlines of Ghana and Nigeria.

The deregulation of the American airlines in 1978 sent ripples through the industry world wide. Fears that there would be increased spatial concentration of activity have not materialised but existing regional cartels could come under threat (Yue-Hong, 1993) and Developing Countries may have to consider the creation of larger operating units if they are to survive under unsubsidised, competitive conditions (Fieler and Goodovitch, 1994). In areas such as the Middle East this may well be possible but smaller African airlines may find it difficult. Most air services are now based on bilateral arrangements which provide the Developing Country with some shelter but a move towards multilaterialism could be disadvantageous for the airlines of smaller countries whose level of development does not allow them to participate fully (Levine, 1993). Only Chile briefly opened up its domestic services to foreign airlines.

Air transport is in some situations a vital ingredient in the development process but it is doubtful if a loss-making airline contributes a great deal even if it is thought to be a reflection, some would think an essential element, of the process of modernisation. However, many smaller Developing Countries may well think that there are also dangers associated with over-dependence on airline services provided externally and over which they have no control – finding an optimal strategy is never going to be easy.

# 5

# ROAD TRANSPORT – THE UNIVERSAL MODE

There can be few if any countries in which road transport is not now the dominant mode by which people and goods are moved and there is a universality about road transport which does not apply to any other mode. In contrast with railways, road transport is characterised by a very wide range of technologies and levels of sophistication. At one end of the spectrum there is the movement of people, or goods carried by people, on foot over some kind of route, possibly of an unimproved kind. Even in the most advanced economies over the shortest distances there is really no alternative to walking. For some this may be no further than to their car in the garage, the local shop or from the railway station to the office. However, for many, indeed possibly the majority, of people in Developing Countries there may be no alternative to walking, whatever the distance involved and irrespective of the fact that goods may have to be carried.

At the other end of the spectrum is the movement of people and goods in specialised road vehicles possibly along engineered roads of considerable sophistication. In advanced consumer-orientated economies, road haulage has assumed overwhelming importance by virtue of its flexibility, convenience and door-to-door capability, whether for the movement of goods or people, yet these same qualities also make road transport a valuable pioneer mechanical mode in economically less advanced areas. Between the head-loading person and the lorry are found a great variety of inter-mediate techologies – including pack animals, hand carts, animal-drawn carts, bicycles, rickshaws and motor cycles.

The 'roads' along which these vehicles move may be no more than temporarily cleared paths, these may be given some perma-nence by the frequent passage of feet or wheels or they may be engineered at varying levels of technology from the graded earth road to the multi-lane express and motorways and the complex, segregated road systems of large urban areas, with pavements and precincts for pedestrians. It is frequently the case that roads and road transport of different levels of technology integrate spatially and functionally into hierarchical systems. In Britain, the 'motorways', class 'A' and 'B' and unclassified roads are very different in engineering and capacity while in Ghana the Accra–Tema motorway, built in the 1960s, parallels an existing road of less sophisticated design through an area where head-loading along un-made routes is still of considerable importance.

Three distinct but closely related elements of road transport change with time. First, there is the develoment of the network. This may be viewed as a series of steps by which an increasing number of nodes are connected and gradually build up into a spatially integrated system of routes with increasing levels of connectivity (Chapter 1). Second, road construction technology has changed with time and third, there is the changing character of the vehicles which use the track. These three elements are inextricably linked, thus, wheeled vehicles necessitated improved surfacing and better alignments. Alignment will be closely related to terrain and will also be influenced by the type of vehicles in use and the road design and construction technology adopted. Thus, improved roads may have to be on a different alignment from the original.

## THE NATURE OF ROAD NETWORKS

In the broad sense in which road transport is here being considered roads come into being as soon as well-defined paths emerge for the movement of people and goods between farmed land and settlements and also between settlements. It is the advent of the wheeled vehicle which invariably prompts road improvement, as illustrated by the development of the Roman strategic road network in Europe, the construction in Britain after 1751 of the turnpike roads and at the present time, the need to construct roads so that motor use can be extended into less accessible areas in Developing Countries.

Figure 5.1, an example from south-east Ghana, represents what might be termed a simple road network. The nodes consist of either single farms or homesteads or small nucleations of the same

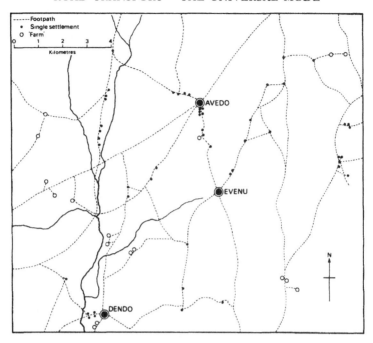

*Figure 5.1* Simple route network – South East Ghana
*Source:* From Ghana Survey, 1:62,500, 1961

and although the overall population density is only modest there is in fact a fairly dense cellular network of paths linking the settlements one with another and with farmed land, sources of water and possibly points with special strategic or sacred and ceremonial significance. In the early 1960s (as represented by the map) farming in this area had not developed much beyond the subsistence level and produce entered into exchange only irregularly and in small quantities and mainly within the local area. The tracks were simply footpaths, were unimproved, and apart possibly from initial bush clearing, were maintained simply by use. In general, paths used by human porters or draught animals were not particularly sensitive to terrain conditions. Although on occasions wheeled vehicles may have ventured along some of the tracks they were not in general suitable for this purpose and in road transport terms they were of low capacity.

It is important to note that at this level of network development there is a high degree of conformity at the micro scale between the route network and the pattern of settlement and its related activities. There is a high level of connectivity in relation to the means of movement (walking, animal transport), new links can be created easily and there is rarely any impediment to a direct move from the track into the surrounding land area at virtually any point. The network has almost infinite fineness of access at the local level. At an early stage in the evolution of any network the links are likely to be uniform in terms of their technology and traffic. Such a largely undifferentiated network of tracks may well survive as long as goods to be carried remain small in quantity and humans the main vehicles of movement and indeed, there was little evidence of change over a very long period of time until quite recently in the road network or the economic activity of the area represented.

The process of upgrading certain routes will be stimulated by demand deriving from the variable growth of the traffic-generating nodes. As the result of a complex interplay of resource endowment, locational characteristics and social, economic and historical forces, settlements come into being, expand, contract and sometimes disappear but invariably combine in hierarchical systems which are dynamic over time. Such a hierarchical structuring of nodes will be reflected in variable traffic-generating capability and an associated differentiation between the links that make up the network. It has already been suggested that even in the relatively early stage described above there is probably some degree of differentiation with respect to the volume of traffic moving on the links making up the network.

As represented in Figure 5.2, the process of differentiation has been taken much further although the area is no more than 20 km from the area already described and is a similar open, savannah-grassland environment. Given the general relationship that has been shown to exist between population and route density (Gould, 1960) it is hardly surprising that the higher population density of this second area is associated with a route pattern of great density and also a more complex cellular structure with a definite hierarchy of route types. Some 50 per cent of the route network is still made up of unimproved footpaths, not even classed as suitable for motor vehicles, and again these conform closely with the pattern of settlement and provide the basic transport infrastructure for an agricultural economy which is still largely self-sufficient and with little entering into trade. One level up in technical terms is what in Ghana is called a 'cycle track' and this

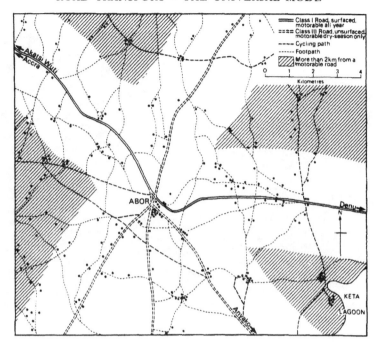

*Figure 5.2*  Upgraded route network – South East Ghana
*Source*:  From Ghana Survey, 1:62,500

represents a marginal improvement on the footpath with possible widening, better clearing but little conscious engineering. The occasional motor vehicle may use such tracks.

The third category, in Ghana the Class III motorable road (dry season only), would typify the 'earth' roads that are a main feature of many Developing Countries. They may be little more than widened paths with the in-situ earth compacted by the passage of vehicles but with only minimal grading and no surfacing. The traffic-bearing characteristics of such earth roads depend on local soil types and drainage in relation to the climatic regime in general and rainfall conditions in particular. The best earth roads are found on well-drained granular soils and the salt-laden soils of 'sabkha' roads in some arid areas (Ellis, 1979) and the 'red-coffee' soils of Kenya can carry considerable traffic flows. There will be a general tendency for wheeled vehicles, especially those with heavy axle loads, to create ruts and progressively lower the road in relation to

161

*Plate 5.1* In most Developing Countries, as in Mali, West Africa, the bicycle is the only form of mechanised transport for many people in town and country.

the surrounding surface. This will lead to rapid waterlogging after rain (Figure 5.3a). Most such roads will become impassable during the prolonged rainy seasons which characterise large parts of the tropical world and vast areas are rendered seasonally inaccessible by widespread road closure. In contrast, these roads will be very dusty in the dry seasons.

The traffic carrying capacity and the rain season condition of earth roads will depend not just on the natural physical conditions but also on the extent to which they have been engineered (Beenhakker *et al.*, 1987). They can be greatly improved simply by raising the road, especially in low-lying areas, clearing the vegetation well back from the road edge to maximise the drying effect of wind and sun and cambering of the surface with side drainage channels to facilitate water run off (Figure 5.3b).

The Class II road in the Ghanaian classification is not represented in the case-study area but would comprise an unsurfaced (in the sense of not having a bitumen or concrete sealing) gravel road, only occasionally closed to traffic. In theory and practice the

162

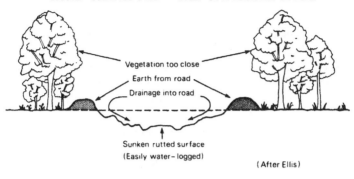

Vegetation too close
Earth from road
Drainage into road

Sunken rutted surface
(Easily water–logged)

(After Ellis)

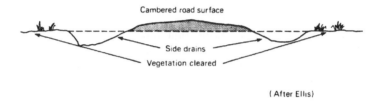

Cambered road surface

Side drains
Vegetation cleared

(After Ellis)

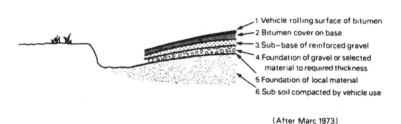

1. Vehicle rolling surface of bitumen
2. Bitumen cover on base
3. Sub-base of reinforced gravel
4. Foundation of gravel or selected material to required thickness
5. Foundation of local material
6. Sub soil compacted by vehicle use

(After Marc 1973)

*Figure 5.3*  Road construction technology
*Source:*  After Ellis (1979) and Marc (1973)

distinction between earth and gravel roads is not always clear and rests mainly on the use, for the latter, of selected, self-binding laterite or gravel material not necessarily found in situ. The widespread distribution in the tropics of lateritic materials suitable for use in road construction tends to blur the distinction. It is often the case that the distinction may also be based on the degree of engineering, with the gravel roads having a better prepared base, some grading, an initial camber and attempts at

163

drainage. The 'swish' roads of West Africa and the 'murram' roads of Uganda (Smith, 1970) utilise lateritic gravel and when newly graded and compacted provide good rolling surfaces. However, such surfaces rapidly deform and corrugate under increasing traffic densities and vehicle weights and continuous maintenance is necessary if the capacity is to be kept at more than about 30 vehicles per day.

When dry these roads are sometimes swept, using a towed branch, to redistribute the surface material. It can be argued that this merely loosens the surface and hastens erosion. During the rains these roads may become impassable and while some engineers believe that this is the most suitable time at which to roll and treat the surface it may be impracticable or even impossible to use the necessary heavy equipment.

As road design becomes more complicated so the terrain exerts an increasing influence and engineering becomes more costly and will therefore be undertaken on an ever more selective basis. It is for this reason that discussion of the evolution of the spatial characteristics of road networks cannot be divorced from technical considerations. Characteristically, Figure 5.2 shows that roads of successively higher levels of engineering tend to take up decreasing proportions of the network and comprise a less cellular and more radial pattern, focussed on major nodes. It is often the case, as illustrated by the Class I surfaced road, that higher-order roads are constructed on new alignments and may even bypass settlements, whereas previous improvements in large measure utilise existing alignments. Motorways and express routes often act more as through routes than local distributors and may bear little or no relationship to the settlement of the area through which they pass. We shall return to this significant point at a later stage.

## ROAD CONSTRUCTION TECHNOLOGY

The example already studied has shown that roads can vary considerably in their design and also in the technology adopted. Indeed, it is because their construction can be phased, that roads in theory are so suitable as a part of the economic development process and there is much greater scope than with railways to adopt design standards and technology that suit existing levels of demand and have the potential for phased improvement as that demand rises.

Pierre Trésaguet has been described as the 'father of modern highway engineering' (O'Flaherty, 1988) and as the Inspector General of Roads in France between 1775 and 1785 he was the first fully to appreciate the vital role of subsoil moisture content and its effect on the stability of the road foundations. The same significance for railroads has already been noted (Chapter 3). He advocated a convex section for both foundation and surface and stressed the need for continuous planned maintenance – a point rarely fully understood even now in many Developing Countries. In early nineteenth-century Britain, Thomas Telford recognised that wear on roads was as much the result of natural conditions (flood, storm, frost) as it was of traffic and he perfected designs utilising broken stones arranged in layers by size and type to ensure good drainage, minimal damage and surface consolidation with the passage of vehicles. The selected gravel roads of many Developing Countries are obviously a step in this direction. He also recognised the significance of alignment and easy gradient.

The best-known name associated with highway engineering is undoubtedly that of John McAdam (1756–1836), who stressed that roads should be made to accommodate the traffic and not vice versa and that two fundamental principles are involved. The first of these is that it is actually the local soil which supports the weight of the traffic and the second is that where this soil is preserved in a dry state it is more likely to do so without subsidence. This necessitates sealing the surface with interlocking stones bound in bitumen to ensure that moisture does not penetrate to the sub base. Thus, whereas Telford emphasised the foundation, McAdam emphasised the surface and the general principles they expounded are as true now as they were two centuries ago.

Possible stages in the improvement of a road surface are shown in Figure 5.3. Starting with the local soil, perhaps already compacted by the passage of some vehicles, a foundation of additional soil can be added and this may be stabilised with the admixture of lime or cement. On this it is then possible to place layers of selected gravel or laterite, compacted, graded and stabilised and finally sealed with bitumen or cement. A simple solution adopted in many Developing Countries is a bitumen spray and chip treatment on a laterite base. The pot holes which characterise so many roads in such areas develop because the soil base is not compacted to a uniform density and when moisture gets into it there is collapse in roughly circular zones. In addition to the gradual improvement of the rolling surface the capacity of roads can also be increased by widening, additional lanes, reduced gradients, better alignments and curvatures, greater sight distances, provision of hard shoulders and improved intersection geometry, all of which affect vehicle speed, operational convenience and safety.

Roads can therefore be categorised according to their design characteristics, their demands on equipment, material and technical expertise and consequently according to their cost. At the one end of the scale they may be cheap, low-capacity earth roads which can be built and maintained with local materials, by local labour with little equipment and minimal expertise and supervision. At the other end of the scale are multi-lane highways in concrete, constructed at great cost and for many Developing Countries with materials, equipment and technical skills that have to be imported. Highway planning is for this reason a crucial area in decision making and while a low cost solution may be desirable the cheapest road is not necessarily the one which will produce maximum returns in relation to investment (O'Flaherty, 1988). The real problem is that of determining which improvements or additions to the network should be made within some given planning horizon and then sequencing the development in an optimal manner (Barber, 1977).

The initial design standard and level of maintenance are closely related and determine the capacity and the cost of using the road. In tropical areas maintenance of low-cost, low-technology roads is extremely important but it is only recently that there has been detailed research quantifying for specific climatic regimes the relationship between the factors involved. Some early work on East Africa (Hodges, Rolt and Jones, 1975) suggested that for gravel roads, vehicle operating costs reflect a complex relationship between longitudinal roughness of the road surface, depth of ruts formed by traffic, amount of gravel lost from the surface as a result of traffic and erosion, the road gradient, the volume of traffic and the amount of rainfall. For any road some maintenance is routine and independent of the traffic volume but much is repair work to damage caused by vehicles. The maximum capacity of a given road may be defined as the traffic density at which it becomes desirable to upgrade or replace it with one of higher capacity and the relationship is shown in Figure 5.4, in which the general cost of

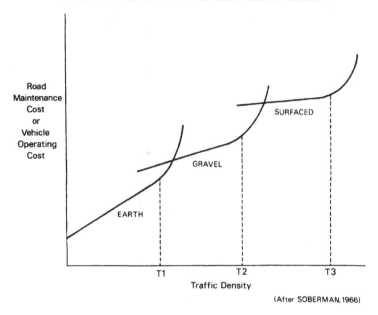

(After SOBERMAN, 1966)

*Figure 5.4*  Traffic density and road transport costs
*Source:*  After Soberman (1966)

either maintenance or vehicle operations are related to traffic density.

For an earth road there will be fairly rapid deterioration of the surface with increased traffic but at a critical density ($T_1$) the cost rises sharply. The actual number of vehicles at which this critical level is reached will depend on surface material, climate, terrain and vehicle types but will probably be in the range of 30 to 50 vehicles per day. In Afghanistan, 100 vehicles a day is accepted as a lower threshold for providing a gravel road (Glaister, 1980). In fact, it is probably more realistic to adopt some measure of standard axle loads and then assess the equivalent number that traverse the road. As a rough guide the damage is proportional to the fourth power of the axle weight – a doubling of the axle weight means 16 times the damage.

When a road is given a gravel surface the initial increase in costs may be less steep than for an earth road but in the same way there will be a critical traffic density ($T_2$) at which there will be a sudden increase in costs. Again the critical level of $T_2$ will depend on a

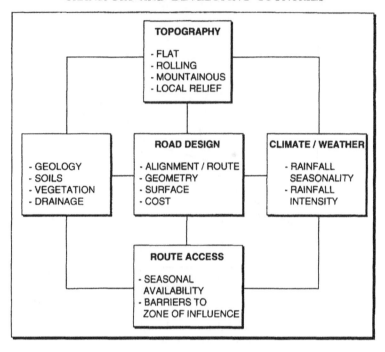

*Figure 5.5*   Environment and rural roads

similar range of variables and in Venezuela a $T_1$ of 50 and $T_2$ of 200 has been adopted, while in Uganda a $T_2$ of 300 has been proposed. In Nigeria a $T_2$ of 50 for a one-lane surfacing and 150 for two lanes has been adopted. There can, clearly, be no simple critical level applicable to all regions but for each environment the identification of the threshold values can provide a valuable guide to the phasing of improvements (Ellis, 1979).

It will be apparent that many of the significant variables are geographical in character (Figure 5.5). Road design criteria are based in the first instance on terrain – flat, rolling and mountainous – with cost for a given standard of road increasing with terrain difficulty (Beenhakker *et al.*, 1987). Likewise, at the broadest level considerations of climate and weather (rainfall quantity and seasonal distribution, frequency and intensity of storms) will influence the design but this will operate at the local level through factors such as slope, soil characteristics and vegetation cover, to influence drainage and erodability. In combination and in relation to the type of road these same factors will influence the availability of the route once constructed and also the facility with which people can approach the route. This will determine the extent of the zone of influence.

## ROAD DESIGN AND CAPACITY

The capacity of roads to accommodate traffic is clearly a function of their design characteristics but there has been relatively little work on the measurement of road capacities in Developing Countries (Boyce et al, 1988). This is not necessarily a problem where roads are functioning at less than capacity and when it is the operating conditions that are of greatest significance. Three stages of road development have been identified with respect to design standards (Kosasih et al, 1987):

- Level 1: Provision of basic access
- Level 2: Provision of increased capacity
- Level 3: Improved operational efficiency and safety

At early stages there may be little attention to geometric standards, it is just the link that is important and whether or not it can be used at all times. With time, capacity considerations will become more important but may well be concerned with the ability of the route to take certain volumes of traffic or whether an improved surface is needed rather than with detailed considerations of geometry except to the extent that they involve road width or gradient. In Papua New Guinea a rural road lane width of 3.25 m is considered technically desirable to allow large lorries to pass but 2.75 m is accepted on economic grounds – an overall width of less than 5.5 m being considered as single lane. On a gravel road of 4.0 m width the capacity of passenger car units per hour is 600 but this increases to 1,900 and 2,400 on roads of 6.0 m and 7.0 m. The effective width is reduced by slow-moving vehicles and the mix of vehicles in many Developing Countries is such that capacity will be greatly reduced. It has been calculated that in India a flow of 100 bicycles an hour in each direction greatly reduces road capacity.

There is clearly a need to provide, especially on narrow roads, occasional widening for ease of driving around curves and also to allow passing and overtaking of vehicles. Again this can be vital where there are slow-moving traditional vehicles, bicycles in large numbers or numerous pedestrians walking on the road and these are best accommodated by the provision of hard shoulders. Given that maintenance of design speed is a critical determinant of capacity it is particularly important that such widening is provided on routes with greater gradients – on non-surfaced roads there is a considerable variation in the traction ability and therefore the speed of two-wheel drive, four-wheel drive and animal-drawn vehicles and also vehicles of different weights.

Gradient is an important consideration but is related to type of road surface and vehicles in use. On gravel roads, gradients of more that 8 per cent can cause problems and as a rule of thumb any road with a gradient in excess of 10 per cent needs to be surfaced. Recommended gradients for different types of roads are shown in Table 5.1. In Papua New Guinea the maximum gradients adopted are 20 per cent for four-wheel drive vehicles, 18 per cent for small goods vehicles, 16 per cent for medium lorries and 12 per cent for timber lorries. Speed reductions on gradients will be most marked where vehicles have low engine capacity in relation to loaded weight or where vehicles have been poorly maintained – both frequently being the case in many Developing Countries.

While proper consideration of design geometry is essential for high-volume roads it is open to debate whether or not the adoption of the same standards is appropriate either technically or economically for what are to be low-volume, and, ideally, low-cost, access routes in rural areas in Developing Countries (Boyce *et al.*, 1988).

*Table 5.1*  Recommended maximum gradient (%) by road type

| | Design speed (mph) | | | | | |
|---|---|---|---|---|---|---|
| | 20 | 30 | 40 | 50 | 60 | 70 |
| a) Local roads/streets | | | | | | |
| level terrain | 8 | 7 | 7 | 6 | 5 | — |
| rolling terrain | 11 | 10 | 9 | 8 | 6 | — |
| mountainous | 16 | 14 | 12 | 10 | — | — |
| b) Rural collectors | | | | | | |
| level terrain | 7 | 7 | 7 | 6 | 5 | 4 |
| rolling terrain | 10 | 9 | 8 | 7 | 6 | 5 |
| mountainous | 12 | 10 | 10 | 9 | 8 | 6 |
| c) Rural main | | | | | | |
| level terrain | — | — | — | 4 | 3 | 3 |
| rolling terrain | — | — | — | 5 | 4 | 4 |
| mountainous | — | — | — | 7 | 6 | 5 |

*Source:* After Kosasih *et al.*, 1987

The fullest consideration of road geometry is less related to issues of basic provision than it is to operational efficiency for the vehicles using the road and involves details of curve radii, sight distances, superelevation (to counteract the sideways force between vehicle and surface), junction layout, gradient and speed. Speed is both the determinant of design geometry and a function of the designs adopted.

The adoption of appropriate design standards becomes critical as concern for safety grows in importance but in many Developing Countries this would not appear to be high on the list of priorities. In Africa, roads are killers of epidemic proportions and in the early 1980s 11 of the top 12 countries for road fatalities were in that continent. Certainly many of the Developing Countries – Nigeria and Thailand are two that are frequently mentioned – have high accident levels and also accidents that are different in kind from most developed countries (Jacobs and Sawyer, 1983). While in part this is related to poor road design standards in relation to traffic volumes it is also a function of the mix of modern and traditional vehicles and pedestrians, poor vehicle maintenance, a general lack of road discipline by all concerned and lack of regulation and proper enforcement.

While the capacity of different types of road will be a function of the design of the road, the condition of the road and the vehicle mix, broad generalised capacities are indicated in Table 5.2.

## ACCESSIBILITY AND DEVELOPMENT

The above discussion and earlier brief examination of specific Ghanaian examples of road development showed that even over short distances there can be considerable variations in the quality of the network and, consequently, in the ease with which people and goods can move about. Figure 5.1 represents a lower order of accessibility, connectivity and convenience than Figure 5.2 and the population has the real disadvantage of having no direct access to motorable roads. At 3.2 km, Denbo is in fact the nearest settlement to a motorable road. Perforce, goods must be headloaded, because this is not an area in which animals are customarily used for draught purposes. It follows that the range of movement and the quantity that can be moved is restricted and the real cost of transport is high.

In the area represented in Figure 5.2 the road network is better

*Table 5.2*  Highway capacities

(a) Theoretical and practical design capacities

| | Theoretical maximum | | Practical design capacity | |
| --- | --- | --- | --- | --- |
| | Ave. speed (mph) | Vehicles per hour | Ave. speed (mph) | Vehicles per hour |
| 2-lane, 2-way | 30–40 | 2,000 | 40–55 | 900 |
| 3-lane, 2-way | " | 4,000 | " | 1,500 |
| 4-lane, 2-way | " | 8,000 | " | 1,000[a] |
| 1-lane max, 1-way | " | 2,000 | " | 1,500[a] |

(b) Capacity in relation to sight distance and vehicle mix

| Terrain | Percentage route with sight distance less than 1,500 feet | Percentage of heavy lorries | | |
| --- | --- | --- | --- | --- |
| | | 0 | 10 | 20 |
| Level | 0 | 870 | 760 | 670 |
| | 40 | 660 | 580 | 510 |
| Rolling | 20 | 770 | 550 | 430 |
| | 60 | 530 | 380 | 300 |
| Mountainous | 40 | 660 | 350 | 240 |
| | 80 | 380 | 200 | 140 |

Design road – two lane, two way, 12 feet lane width, 50 mph

*Source:* (a) Hay (1961); (b) Paquette *et al.* (1982)
*Notes:* (a) in direction of heaviest flow.

developed and, ignoring the complications which arise from the arbitrary selection of the area, there are only small zones which are more than 2 km from a motorable track. Certain authorities, including the World Bank, adopt 5 km as an acceptable distance beyond which people should not have to walk to a motorable route but even this is arbitrary and its validity dependent on whether head-loading or animal transport is used. Research has suggested that where there are no physical barriers to pedestrian movement, the zone of influence of a road is only 2–3 km, although for practical reasons this may have to be increased to 6–12 km (Beenhakker *et al.*, 1987)

On this basis, large parts of Developing Countries are found to be inaccessible and by general standards the Ghanaian example represents a relatively good level of connectivity, access to markets for smaller quantities of agricultural produce and access to medical, educational or other services will be comparatively easy. Over parts of Mexico it has been estimated that 70 per cent of the population has only foot or horse transport available (Edmonds, 1980) and in the 1970s there were 12 million people in 16,000 villages not within an acceptable distance of year-round motorable roads. In Ethiopia at the same time 45 per cent of the population was more than 16 km from an all-season road (Baker, 1974) and the situation has changed only little since that time.

In Chapter 1 it was suggested that two of the main characteristics of networks are connectivity (the amount of linkage between nodes) and fineness (the number of points at which one has access to the network) and in general, on both counts, road networks score highly in comparison with railways and waterways – railway stations and river ports are usually few in number and very widely dispersed and therefore offer little hope of providing the access required by scattered, mainly rural populations. It is for this reason that roads have such a special role in the development process. This is a point which did not escape the attention of colonial authorities and in many Developing Countries the highly selective network of improved roads has changed little from that established in the colonial era and gives rise to marked regional disparities in the levels of accessibility enjoyed by the population. Smith's (1970) study of the Ugandan network makes this point most forcibly and similar conclusions derive from work in Malaysia (Leinbach, 1975) and Sierra Leone (Sessay, 1968). In India in the 1970s 32.5 per cent of the surfaced roads were concentrated in only 12.5 per cent of the national area (Gananathan, 1973) and in Jamaica 41 per cent of the road mileage variance has been attributed to the location of sugar plantations, which were markedly coastal in location (Hubbard, 1974). These regional inequalities have become a main target of road improvement programmes in recent years but are still far from being eliminated.

All too frequently inadequacies of road transport are cited in explanation of poor economic performance (e.g. evacuation of Ghana's cocoa crop has often been held up) and lack of road capacity becomes apparent when there is the need to transport large quantities of, for example, relief aid into remoter areas of the Sahel, Karamoja, Ethiopia or Kampuchea. These same examples also illustrate the point that the efficiency and adequacy of road transport is a function not only of the network itself but also of vehicle availability and suitability for particular tasks in specific conditions. Roads are of no use without road vehicles.

In many of the Developing Countries there was scant attention to the development of a road network which would serve the bulk of the population and it is only recently, with a better understanding of the development process and greater emphasis on agriculture and rural populations that a more critical approach to road construction has been adopted and some attempt made to remedy the regional inequalities already noted. Indeed, as Mabogunje (1980) has argued, spatial restructuring must be an essential element in any meaningful development strategy and given their all-pervasive influence it is clearly by way of road provision and improvement that this will be achieved.

The close relationship between settlement and route patterns has already been noted and McMaster (1970b) suggested that the influence of roads on development operates largely through their impact on settlement. It is undoubtedly true that in many places settlement patterns have adjusted to new roads and there has often been considerable growth of population at strategic points in evolving networks – the attraction of people, services and economic actvities to cross roads provides an obvious example. Garrison (1959) showed how increasing mobility resulted in a re-patterning of settlement and service provision in the United States and the process of urbanisation, not necessarily good in itself, is certainly assisted by improved roads and in rural areas the nucleation of settlement will be encouraged. The actual influence of a road will reflect its design standards, geographical factors (e.g. seasonal availability related to climate) and the ease with which people can reach it. It will also reflect the resources of the area and the range of factors influencing demand (Chapter 1).

In the 1960s the Ghana government constructed a multi-lane, motorway standard road of 24 km in length to link the outskirts of Accra, the capital, with Tema, its main port. This was a toll road with restricted access and with wide hard shoulders, sophisticated drainage engineering and was, for obvious safety reasons, fenced off from the surrounding area. Such a road closely resembles a railway line in that, at the local level, it tends to disrupt rather than integrate movement and economic activity and any positive impact will be restricted to the terminal points and, where provided, any intermediate access points – no such points were provided along the Accra–Tema road. Such a road will have its maximum benefit, perhaps its only benefit, at a regional scale, few nodes will be affected and there is little or no chance of spontaneous attraction to the routeway and the emergence of new nucleated settlement.

In contrast, as we have already seen, the seasonally motorable

174

*Table 5.3*  Road types and development impact

| Road type | Relationship to settlement and economy | | Relationship to emergence and growth of new nucleated settlements |
|---|---|---|---|
| | *At regional level* | *At local level* | |
| 1 Motorway, restricted access | Maximal inter-urban impact | Negative | Nil |
| 2 All-weather surfaced | Maximal inter-regional, access to ports, industrialisation | Average and decreasing, reduced effect in dissected country | Considerable growth at major nodes and intersections |
| 3 All-weather gravel | Moderate, feeders to surfaced roads | Average | Maximal emergence of new lower-order central places |
| 4 Dry season roads | Slight | Considerable | Slight |
| 5 Motorable tracks | Minimal/nil | Maximal, accord with maximum persons in rural areas | Minimal |

*Source:* (Modified from McMaster, 1970)

track has almost infinite fineness at the micro level and can accord with even the most dispersed settlement pattern. It can be used by people, animals and by a wide range of vehicles and will be the means by which the maximum number of settlements and people can be given a degree of accessibility. Between the two extremes just described are road types with differing degrees of fineness and accord with rural settlement. McMaster (1970b) has proposed (modified in Table 5.3) a relationship between road technology and the type, extent and degree of spontaneity of social and economic response. McMaster cites the example of eastern Nigeria, where nucleation is the typical settlement pattern and where there has been a strong tendency for new settlements to emerge along improved roads. In contrast, in the banana-growing area of East Mongo, Uganda, there was a close accord between the dispersed settlement and motorable tracks but little evidence of relocation or the emergence of new settlements on roads improved to a surfaced, all-weather standard.

It is wrong to assume that roads will automatically lead to increased productive activity and in some cases, even in now independent territories, they may be provided for social, administrative or overtly political reasons. An improved road may be no more than an expression of political influence or a reward for services rendered but it is always the case that 'the highway is simply a functional element in the area plan and thus its justification, location and quality cannot simply be considered in the light of strictly economic considerations' (O'Flaherty, 1988). If, as is now increasingly asserted, it is the rural areas that hold the key to many development problems, then roads must be built in greater numbers to provide the minimal levels of accessibility on which development, in the broadest sense, must be based. It has also been claimed that in many Developing Countries it is at the level of the 'smallest branches of the network tree, the low-volume feeder roads, tracks, trails and paths that join the rural population to the roads' (Beenhakker et al., 1987) that the weakest links are to be found and it is the rural feeder roads which must be given priority in any meaningful development strategy. This inevitably brings us to the question of low-cost roads.

## LOW-COST ROADS FOR DEVELOPMENT

Given the vast areas and numbers of people effectively excluded from the development process by inaccessibility it follows that the necessary road networks will have to be provided at the lowest possible cost and therefore at a level of technology which, while adequate to satisfy demand, will maximise the use of available resources, human and material, and minimise the use of scarce resources such as capital and technical expertise which may have to come from outside.

In some cases the roads will be provided as an integral part of specific agro-industrial development projects. Perrusset (1977) described how, in Gabon, forestry companies developed road systems for the evacuation of timber from their concessions. The success of the Mumias Sugar Company in western Kenya prompted the establishment of other projects at Kakamega (1976), Siaya (1977), Nzoia (1978) and Sony (1980) and some 50,000 smallholders, until that time mainly operating at a subsistence level, started to grow the cane sugar. At Mumia, 17,000 farmers within 21 km of the mill each have 1.2 to 2.0 ha of cane

*Plate 5.2* Road construction and repairs, as in Bangladesh, may be locally organised and labour intensive, but earth and gravel roads deteriorate rapidly under heavy vehicle use and may be seasonally impassable (Photo: Intermediate Technology/Neil Cooper).

and all services are provided on credit by the company. Each scheme has its own feeder road system and this amounts to over 3,000 km of access and 650 km of feeder road linking farmers to the mills. This is in addition to 400 km of upgraded existing road, mainly to bitumen-surfaced standard. In order to ensure the continuous maintenance so necessary for heavily used roads in tropical regions, each mill has been provided with a road maintenance unit which assumes responsibility for all the roads in the project area. Also in Kenya, a Rural Access Roads Programme, started in 1974, was to provide 14,000 km of roads by labour-intensive construction methods, with casual employment providing between 2,000 and 20,000 jobs at different times. In 1986 Nigeria's Directorate of Food, Roads and Rural Reconstruction (DFRRI) was created to mobilise the stagnant rural sector and, to date, over 61,000 km of rural roads have been constructed or rehabilitated (Filani, 1993).

The extent to which labour can replace machinery in road construction depends on the need for earthworks and structures such as bridges and culverts, the road design and surfacing and the terrain. For example, the gradient adopted dictates the amount of earthwork required in any particular terrain. Likewise, if a design speed for a road allows it to follow easily the contours this will reduce earthworks and the need to employ heavy equipment. As the Chinese seem frequently to demonstrate, even very considerable earth moving is possible with intensive labour methods but a road geometry can be adopted to minimise the effort required.

Starting in the early 1970s the Mexican government embarked on a scheme using labour-based construction methods to provide road access for the mass of the rural population (Edmonds, 1980). The starting point in each case was to be a request from a village for a new road or the upgrading of an existing road. The actual need in relation to numbers of households, social services and agricultural development potential was then assessed and a committee set up, including people from the community and district engineer's office who were to oversee the project. The engineer's office provided technical expertise, material needed, hand tools and equipment, transport and payment for the labourers being provided by the community The actual routes were chosen on the basis of maximising the route utility in terms of the number of settlements served, the existing socio-economic infrastructure, the potential for agricultural development and employment generation and the zone of influence of the completed road.

The standard adopted may be classed as 'low technology' with a design speed of only 20–30 kph, curve radii of as little as 15–20 m and a general gradient not more than 12 per cent. The surfaces were 4.0 m of gravel, except where grades had to exceed 12 per cent, where cobbles were used, with widened passing places at every 500 m to allow 50 vehicles a day on what was basically a one-way flow. The employment generated amounted to an average of 1,700 man-days per kilometre, but varied with terrain conditions. During the 1970s over 64,000 km of such roads were completed, involving improved access to over 6,000 settlements. The accessibility of many of the communities with respect to schooling, medical services, markets and the time to the nearest town has been greatly improved. Nevertheless, in the absence of any rural development programme there was the tendency for existing economic structures to be reinforced, and with the better off gaining most benefit and the additional income generated being spent in the towns rather than in the rural areas (Edmonds, 1980).

In Afghanistan (Glaister, 1980), in recognition of the advantages of improved transport, the Rural Development Department initiated self-help road construction projects with designs adapted so that unskilled labour could be used and imported materials and skills virtually eliminated. The basic design was for roads able to take 100 vehicles a day with a mix of 17 per cent cars/jeeps, 33 per cent light trucks and buses and 50 per cent medium trucks and buses. The 3.5 m roads were widened to 6.5 m at 500 m passing points and an inside bend radius of only 9.0 m was adopted. Where low levels of traffic were anticipated it was assumed that delays of 24 to 48 hours would be acceptable and this determined where it would be necessary to build bridges rather than use fords. Gravel surfacing was used for roads with higher levels of use but for the most part earth was used. The overall gradients adopted were 3 per cent on earth and 4 per cent on gravel with outside maxima of 10 and 12 per cent respectively. In the difficult terrain which characterises much of the country the final alignments of roads, and especially the location of bridges, must, as far as possible, be determined at the outset and where final alignments cannot be fixed with certainty it was thought wise to use only temporary structures (e.g. Bailey type bridges). After years of conflict it is not known what the final outcome of this programme was but it is an interesting example of a conscious attempt to phase construction in relation to demand.

## PLANNED ROAD IMPROVEMENT

With roads as in other areas of infrastructure provision, the Developing Country should be attempting to gain maximum impact for minimum cost and this requires the careful identification of goals and the assessment of demand, revealed and latent (Chapter 1). It is therefore a question of providing access for specific purposes.

However, as Beenhakker *et al.* suggest (1987), access may be at different levels. *Minimum access* could imply that it is not continuous and perhaps effected only with difficulty and this is the level now available for many communities in the Developing Countries. There may well be some scope for low cost improvement. *Reliable access* may involve a degree of closure of the route as long as this

*Table 5.4* Sufficiency rating in road design

| | *Maximum points* |
|---|---|
| Structural adequacy | |
| subgrade | 10 |
| base | 20 |
| surface | 12 |
| drainage | 8 |
| Safety | |
| surface width | 7 |
| shoulder width | 8 |
| stopping sight distance | 10 |
| alignment consistency | 5 |
| Service | |
| alignment | 5 |
| passing sight distance | 5 |
| surface width | 5 |
| rideability | 5 |

*Source:* After Howe (1984)

does not mean an unacceptable constraint on the desired social and economic activities of the particular community – and what is considered reliable in one place may not necessarily be so considered in another. Finally, there is *continuous access* in which there are no constraints on transport and travel but over-design must be avoided and design related to local circumstances.

One possible approach is to assess the adequacy of the existing route by rating its sufficiency in various respects. Points can be allocated to a maximum assigned to each characteristic, this weighting being arbitrary in nature and the whole representing an emphasis on the engineering features of the road (Table 5.4). However, there is no reason why such a sufficiency rating should not include non-engineering elements. The ratings provide a basis for effecting minor road improvements or establishing a maintenance schedule (Howe, 1984). It is but a short step to an assessment of the relative costs and benefits of the range of strategies available, whether at the level of selecting between alternative minor improvements to existing roads, choosing between different roads for improvement, or determining the technology and construction methods to be adopted (e.g. labour-intensive or machine-based).

A valuable guide to the planning and improvement of rural roads has been provided by Beenhakker *et al.* (1987). Their starting point is an acceptance of the critical role of access, this being the ability to transport production surpluses to external markets, receive imported goods and services, engage in internal trade, move between settlement and natural resources, transport the essentials of everyday life and ensure personal mobility for economic and social activities. Access is a function of infrastructure (paths, tracks, roads) and transport aids, namely the variety of means both traditional (baskets, carts, animals) and conventional (lorries, buses) by which goods and people are moved. These must always be considered simultaneously and will combine in different ways to satisfy each type of demand. They argue that standardised solutions are unlikely to be effective and they outline procedures for identifying the transport problems that exist in particular situations and strategies for their solution.

Access demand has to be evaluated in terms of the seasonal pattern of productive activity whether in farming, fishing or forestry and minimum and reliable levels of provision established. The selection of transport aids will be determined by availability and suitability for specific purposes, the identification and solution of constraints on their use (technical, institutional, user) and the investments required. The infrastructure provided will depend on assessments of demand and aids, the level of access thought to be appropriate and the finance available. Beenhakker *et al.* argue with justification that in many Developing Countries lack of maintenance of roads, and indeed other infrastructure, is often seriously neglected and it is essential that maintenance programmes are planned and financed as an integral part of the package of transport provision. In Ghana in the 1980s there were mounting problems of transporting cocoa from producing areas to ports because road surfaces were in such a bad state – a combined function of too many unsuitable vehicles on inadequate road surfaces for which there had been little maintenance. The problem was aggravated by the age and state of the transport aids – the lorries.

## THE CONSEQUENCES OF ROAD IMPROVEMENT

Some attention was given in Chapter 1 to the general question of transport and its impact on development but certain more specific examples relating to roads are now appropriate. What exactly are the effects of new or improved road transport?

The consequences of road transport are multiple and, in the long term, all-pervasive, and a useful attempt at classification was presented earlier in this book (Table 1.4). Some of the consequences will relate directly to the user of the new or improved road but others indirectly to the immediate zone of influence or to the wider region or even nation. Some of the influences will be quantifiable in market or monetary terms but others may be less readily evaluated.

It does not follow that the consequences are necessarily advantageous or desirable and many are certainly not assured and may well have to be counted as costs rather than benefits (Porter, 1995). There is the complication that the monetary benefits or costs of particular consequences will be unevenly distributed both over time and also over space and it may well be a very long time before any realistic total balance sheet can be drawn up and the full consequences assessed. It follows that 'before' and 'after' studies may at best give only a partial indication of trends and are unlikely to provide a complete or final evaluation of the project concerned – the time spans adopted are invariably too short, the relevant time-series statistics are either lacking or inaccurate and the data is often aggregated in a way which makes judgements about long-term effects of doubtful validity (Hofmeier, 1972).

Some examples will serve to indicate the variety and scale of the consequences of specific projects. In the 1950s the road from Dar-es-Salaam to Morogoro, Tanzania, was constructed on a completely new alignment through sparsely populated country to the north of the existing railway line which it paralleled. By the late 1960s there had been considerable expansion of cultivation along the line of the new road, the decline or even disappearance of cultivation along the original road and a decrease in local traffic moving on the railway (Hofmeier, 1972). There had been spontaneous movement of population to the roadside and surveys indicated that people were particularly eager to gain better access to medical facilities and markets for their produce. Hofmeier demonstrated that traffic increases along new roads could be quite spectacular (Figure 5.6) and that while this was in part a diversion of traffic from poorer, existing routes it also included considerable elements of newly generated traffic.

Ward's study of the Rigo road in Papua New Guinea provided similar evidence of rapid traffic generation but also pointed to the lack of official statistics (Ward, 1970) and the problems associated

*Plate 5.3* Only selected routes can be improved to higher standards and their influence may be regional rather than local and they may bypass smaller settlements (Photo: Julius Berger).

with studies of road projects. Nevertheless, the 61-km road from Port Moresby to Rigo, upgraded from a rough, seasonal track to gravel, all-weather road in the mid-1960s brought an almost immediate fivefold increase in vehicle movements and there was new vehicle ownership in settlements with no previous access to the road (suggesting the value-adding effect of feeder roads). Even without official encouragement or support there was a marked expansion in village gardening although there were no short-term changes in the actual methods employed. Greatly increased quantities of surplus production found their way into Port Moresby markets and Ward estimated that over a three-year period the net sale value was in excess of the cost of the road improvement. Ward demonstrated reduced freight transport costs and real improvements in convenience, certainty and time saving and confirmed Hofmeier's general conclusion that 'the new construction or the improvement of a road in an area where all the other necessary prerequisites are already existing can significantly encourage and

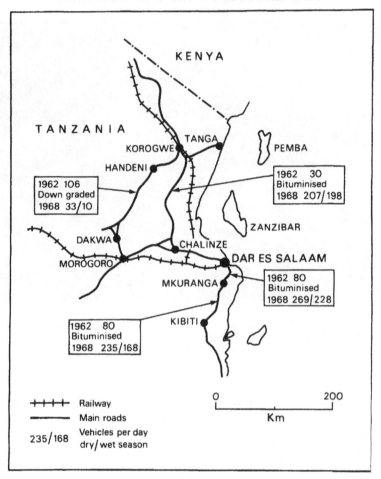

*Figure 5.6* Changes in road use, Tanzania
*Source*: After Hofmeier (1972)

accelerate economic development' (Hofmeier, 1972). Yet as Howe (1984) cautions, it is wise not to read too much into such claims and he cites a study by Bouchard of a road development in the Okapa region of Papua New Guinea where there was little evidence of much change in marketable production.

The significant point to emphasise is that the road will not inevitably lead to development except in situations where transport is the only obvious missing factor. A study of Guatemala's

Atlantic Highway (Klein, 1966) showed that a road designed and constructed in haste, poorly maintained on completion and paralleled over its whole route by an existing railway had little impact. Its potential was latent and dependent on investment in other sectors such as settlement, power, drainage, irrigation, health and education. Yet there is little evidence that such well-chronicled failures had any real impact on some later developments – Ghana's motorway has already been cited.

## AMAZONIAN HIGHWAYS

Although they can be dealt with only briefly here, the roads constructed in Brazil's Amazonian demographic void have generated considerable research and literature (Kleinpenning, 1971, 1978; Goodland and Irwin, 1975; Smith, 1976).

In 1964 a new surfaced road was completed between Brasilia and Belem through a region previously largely inaccessible except by air and river transport. The initial impact of this road was spectacular and has been summarised by Resende (1973) (Table 5.5). It can, of course, be argued that the environmental impact of this, like many of the later roads in the region, was disastrous but it undoubtedly had the effect of attracting to it a considerable population, it resulted in the emergence of new central places, it stimulated agricultural activity, it greatly increased accessibility to large areas and traffic generation exceeded all expectations.

*Table 5.5*  Initial impact of Brasilia–Belem highway

|   | | 1960 | 1970 |
|---|---|---|---|
| 1 | Population along route excluding Brasilia, Belem and Anapolis | 100,000 | 2,000,000 |
| 2 | Number of settlements | 10 | 120 |
| 3 | Cultivation | Subsistence | Intensive cultivation of maize, beans, rice, cotton |
| 4 | Livestock (cattle) | Unknown | 5,000,000 |
| 5 | Vehicular traffic per day | Almost nil | Anapolis – Uruacu 700 Uruacu – Guripi 350 Guripi – Belem 300 |
| 6 | Side roads | None | 2,300 km |

*Source:* After Resende (1973)

It was clearly in the hope that such 'successes' might be repeated that the Brazilian government in the 1970s embarked on the 5,400-km Transamazonica Highway (Figure 5.7) from the Atlantic coast at Recife in the east to the border with Peru in the west. There would appear to have been no initial environmental appraisal (Goodland and Irwin, 1975) and approval and start of construction were hasty in the extreme. The severe drought of 1970 in Brazil's over-populated North-East region certainly provided a spur to the road scheme which was routed to traverse areas thought suitable for cultivation and animal husbandry and hence for colonisation from the North-East. In addition there were strategic and security considerations and the more general aim of improving national integration.

The road is across undulating topography and consists of a 70-m cleared strip with a 8.6-m pavement width mainly of local gravel. Drainage is provided by culverts and side channels and rivers of less than 100 m are spanned by wooden bridges and those of greater width by means of self-propelled raft ferries. The road crosses the heads of navigation on the Araguaia, Xingu and

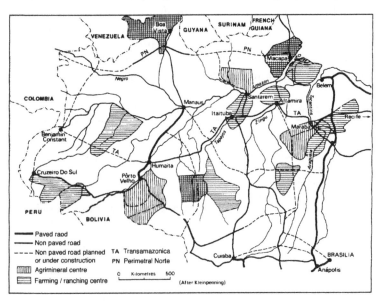

*Figure 5.7*   The Amazon highways
*Source:*   After Kleinpenning (1971, 1978)

Tapajos rivers and at these and many other points integration of road and river transport is possible.

A 100-km zone either side of the road was expropriated by the government to avoid speculation and a 20-km zone was divided into 100-ha lots for agricultural colonisation. These lots were found to be of variable soil quality and agricultural potential and the rigidly rectangular survey means that feeder roads are geometrical rather than adapted to the terrain and this reduces their use (Wesche, 1974). It is also the case that the main road tends to follow the terra firma which may be better for road construction but is less suitable for agriculture than the terra rosa. The rate of colonisation was slower than predicted, certainly by smallholders, and there would seem to have been a reluctance to move into the smaller agrovilas, in part because of the poor quality of the feeder roads and the associated lack of short-distance public transport and private vehicles. It became policy to concentrate attention on the larger agropoles (1,500–4,000 population) and ruropoles (20,000) with smaller settlements stagnating and increasingly large-scale clearance for cattle ranching became the norm.

There would appear to be a general feeling that the whole project was embarked on too hastily, with an unthinking development of land irrespective of soil quality, lack of selectivity in clearance and 'reckless occupation' (Wesche, 1974), undoubtedly speeded up since that description was applied, which is resulting in disastrous environmental and social consequences. It may be that final judgement is premature since much of the planned road network has still to be completed, many of the proposed agricultural development schemes are still in their infancy, some are likely to be stillborn, and many of the envisaged mineral exploitation schemes are no further than the drawing board. The original plan to provide homes and livelihood for the poor of the heavily congested North-East has been replaced by plans for much larger-scale exploitation of agricultural, forestry, mineral and energy resources in selected regions and rapid development of the primary sector with a likely polarisation of economic effort and benefit in both spatial and social terms.

In turn, this new emphasis on large-scale developments could well require an upgrading of the planned road network. The road surfaces will be unsatisfactory for really heavy traffic and the numerous water courses with great fluctuations in seasonal flow could create serious disruptions to traffic unless properly bridged or provided with high capacity ferries. As traffic builds up and vehicle sizes increase many of the crossing points may need improvement.

## OTHER ROAD DEVELOPMENT SCHEMES

In some ways comparable in scale with the Transamazonica but with an emphasis on international integration and trade as well as economic development along its route is the Trans Africa Highway project. This has been on the drawing board for many years and in part utilises existing roads of variable but mainly very poor quality which are to be upgraded to a uniform 7.3-m pavement with a 100-kph running surface. Progress on the highway has been left to individual countries and with more urgent problems progress has been patchy. Over long distances the route traverses negative terrain and it is difficult to identify any substantial, short-term traffic generating potential.

Given the assumed significance of roads and improved access for rural development it is not surprising that many Developing Countries have embarked on rural road projects of varying scales, technologies and administrative and organisational style. In general there has been a considerable difference between stated intentions and actual implementation.

In India, there has been road planning since 1943 when a formula was established by which each state could assess its road requirements with the end objective that no village would be more than 3.2 km from a motorable road (Thomas, 1984). This was followed by a succession of rural road plans but still many smaller villages have inadequate access and the 1981–2001 plan drew particular attention to the urgent needs of rural, hilly, tribal and undeveloped areas but also argued a need for full consideration of non-mechanised transport. Many of the roads are unsurfaced and in effect little more than paths. The increase in food grain production from 55 million tonnes in 1950 to 150 million tonnes in 1985 was initially the result of an expansion of the area under cultivation but with time depended more on application of greater quantities of fertilisers and pesticides to high yielding crop varieties. The most accessible areas were able to benefit most as even drilling equipment for tube wells requires improved transport (Owen, 1987). A World Bank-sponsored programme attempted to improve milk collection from farms and enable more small-scale producers to participate. Two million farmers in some 10,000 villages now produce milk, some of them with few cattle and in very small quantities, for Delhi and other large cities. In this case the improved transport was part of a package of inputs which included participation by large milk-producing companies and technical and financial assistance to the farmers.

Howe (1984) has described the problems of rural access in Botswana where the overall population density is very low, pastoral farming the predominant activity and areas suitable for arable farming restricted. A rural roads programme was initiated in 1973 with the object of using labour-intensive technology. However praiseworthy, labour-intensive methods were soon abandoned, possibly because expatriate staff were not fully committed, and some roads were possibly built to higher standard and at greater cost than was really necessary. This led to the suggestion that it was improvement rather than new road construction that was needed and that selection of routes for improvement, always a difficult and sometimes politically influenced issue, was not always as rational as it should have been. While social considerations may be involved at early stages of discussion in general it is fairly standard economic evaluation which dominates the selection procedure and this does not always produce maximum development benefit.

Bolivia, Kenya, Nepal, Tanzania, Thailand and Zimbabwe are a few of the many Developing Countries in which there have been rural road programmes but even these differ greatly in their scale and the manner in which they are managed. In Zimbabwe, over half of the population lives in the pre-independence defined Tribal Trust Lands but after independence in 1980 there was a purposeful attempt to improve access in these mainly agricultural areas of out migration and land degradation (Hoyle and Smith, 1992). In 1984 a Rural Roads Programme was initiated under which 16,000 km of all-weather primary roads suitable for bus transport were to be built so that all households would be within 10 km – this being determined by a day's return journey for the animal drawn carts in common use. A system of seasonal secondary roads brought home-to-road distances down to five kilometres and in hilly areas where animal carts were impracticable the primary road density was increased to reduce this to a maxiumum of three kilometres. The impact of the new roads has been considerable with much improved access, more journeys over longer distances and an expansion of

*Table 5.6*   Thai Accelerated Rural Development roads

|  | Width (m) | Laterite depth (cm) | Maintenance |
|---|---|---|---|
| Standard ARD road | 6 | 20 | Province |
| Village Access I | 5 | 15 | Province |
| Village Access II | 3–5 | 10 | Village |

*Source:* Based on Jones (1984)

market areas for rural crops and a shift from a largely subsistence to a more cash-oriented economy (Smith, 1989).

In Thailand, the period 1959 to 1977 saw a dramatic increase in road provision both at the primary, national/regional and secondary, local level and by 1981 there were 44,000 km of main and 86,000 km of secondary roads, with only remoter western margins along the Myanmar border being inadequately served (Jones, 1984). The government's Department of Highways is responsible for the national and provincial links, some 45 per cent of which are now surfaced, and the Office for Accelerated Rural Development (ARD) has been responsible for expanding the rural roads, all of which are laterite surfaced (Table 5.6). By 1981, 17,000 km of ARD road had been built and a further 60,000 km were under consideration with an arrangement whereby, when traffic reached 300 vehicles a day and maintenance costs became excessive, the responsibility switched to the Department of Highways, which would then surface the road. The selection of routes for improvement or new road provision is consciously phased and based on an allocation of points for the number of people who would benefit directly, number of people within a 3-km zone either side of the road, current agricultural production, likely increases in cultivated area, current passenger demand and estimated construction costs.

Thailand has experienced sustained economic growth over the last several decades with average per capita incomes rising and a declining and small proportion of its population in the absolute poverty group. There has been a growing contribution of the rural areas to this growth and improved road provision is undoubtedly one of the main factors involved. However, if Thailand's road network is above average by Developing Country standards and its rural areas have been brought into the development process it is finding it increasingly difficult to sustain industrial expansion as it becomes necessary to locate new industries away from the heavily congested Bangkok area (Owen, 1987).

## VEHICLES FOR ROADS

An almost universal feature of road transport has been its rapid increase in recent years, a reflection of its flexibility and door-to-door convenience. Some measure of this increase is given in Table 5.7. High percentage growth in the vehicle numbers for many Developing Countries represents low absolute starting points and private car ownership levels are still in most cases extremely low. The number of commercial vehicles is a reflection of the character and distribution of economic activity and the availability of other forms of transport. Rapidly developing countries such as Brazil, or at a lower level Nigeria, show high growth rates and often the existence of large private or state-owned transport enterprises.

The flexibility of road transport derives in part from the density and scope for phased development of the network but also in the great variety of possible vehicles in terms of capacity, levels of technology and propulsion. It is a feature of the road haulage

*Table 5.7*  Growth of world vehicle fleet (1,000s)

|  | Passenger cars | | | Commercial vehicles | | |
|---|---|---|---|---|---|---|
|  | 1951 | 1971 | 1990 | 1951 | 1971 | 1990 |
| World | 58,880 | 205,630 | 443,430 | 16,190 | 53,960 | 122,261 |
| North America | 43,250 | 101,890 | 156,173 | 10,240 | 21,820 | 49,093 |
|  | (79.7) | (49.6) | (35.2) | (63.2) | (40.4) | (40.1) |
| Europe | 7,000 | 73,200 | 165,879 | 3,550 | 11,420 | 22,249 |
|  | (12.5) | (33.5) | (37.4) | (21.9) | (21.2) | (18.2) |
| Oceania | 1,270 | 5,110 | 8,960 | 670 | 1,200 | 2,293 |
| Asia | 580 | 14,450 | 52,098 | 760 | 11,450 | 33,441 |
|  | (1.0) | (7.0) | (11.7) | (5.8) | (21.2) | (27.4) |
| Japan | 58 | 10,570 | 34,924 | 341 | 9,097 | 21,571 |
| India | 159 | 671 | 2,790 | 120 | 562 | 3,434 |
| Central and | 890 | 5,740 | 29,801 | 680 | 2,250 | 10,678 |
| South America | (1.6) | (2.8) | (6.7) | (5.2) | (4.2) | (8.7) |
| Brazil | 316 | 2,787 | 10,598 | 226 | 687 | 2,472 |
| Africa | 640 | 3,540 | 9,072 | 290 | 1,420 | 4,262 |
|  | (1.1) | (1.7) | (2.0) | (2.2) | (2.4) | (3.5) |

*Source*: Calculated from UN Statistical Yearbooks
*Notes*: (79.7) – percentage of world total

industry that, unlike other forms of transport, it forms a significant sector for local entrepreneurial activity and small-scale investment. For example, in parts of Africa it is quite normal for successful farmers or traders (often women as well as men and hence the term 'mammy lorry' in West Africa) to invest in a 'taxi' or 'lorry' – often to be driven by a family member and suitably inscribed with slogans which make the point e.g. ENAM OBI SO on a Ghanaian lorry – 'it comes from somebody'. It is also the case that investment may be by kinship groups, cooperatives, village or church groups, schools or small businesses.

A great deal of the transport in Developing Countries is of a relatively 'informal' type in which vehicles may wholly or in part be built locally to suit conditions of local environment and market. The demand for transport in many Developing Countries is not as specialised as in more advanced economies and vehicles with variable passenger/freight carrying configurations are needed. Services operated by such vehicles operate largely by inducement rather than to schedule and therefore adapt readily to changing demand. While sometimes in competition with other modes, vehicles of this type provide an ideal medium for meeting latent demand in inaccessible areas not served by other transport.

While impossible to quantify precisely, it is undoubtedly the case that much of the transport in Developing Countries is effected by even less formal vehicles. In many rural areas human porterage is still the dominant mode – e.g. 'head-loading' in Africa – or possibly by hand- or animal-drawn cart. Increasing levels of sophistication are provided by bicycles, bicycle trailers and tricycles or motorised versions of them.

In India most of the farmers with less than 5 ha of land and 74 per cent of all households do not possess any form of wheeled vehicle; bullock carts remain the main form of transport and in many rural areas motorised vehicles are still a rarity – animal carts account for nearly 70 per cent of the tonne-km performed, with head-loading accounting for about 10 per cent. Most of the animals used in transport are also used in agriculture and the marginal cost for transport is therefore low or even zero (Barwell *et al.*, 1985). In Nigeria in 1981, 32 per cent of the households had an animal for transport purposes and 68 per cent had a bicycle (Hathway, 1985).

There can be no doubting the overwhelming importance of low-cost, low-technology, traditional vehicles in providing for a high proportion of the mobility in many Developing Countries and there is no likelihood that this will change significantly in the near future. Indeed, the capacity of the vehicles and technology employed are closely related to local conditions of environment, economy and demand and are very much a factor in the flexibility of road transport in relation to development requirements. Some feel that these informal modes could play an even more significant part and that their capacity and efficiency could be greatly increased by simple and relatively inexpensive improvements in technology which have the advantage of being very much within the capability of local craftsmen and undemanding of expensive imports (Hathway, 1985).

Eight distinct categories of low-cost vehicle have been identified (Hathway, 1985):

1 carrying aids for porters;
2 wheel barrows and handcarts;
3 animal transport;
4 pedal-powered vehicles;
5 motorised cycles and conversions;
6 cycle and motor cycle trailers;
7 basic motor vehicles; and
8 agricultural vehicles.

These vehicles vary greatly in their cost, capacity, speed, range and route restrictions (Table 5.8) but there is in many places scope either for moving up the technical hierarchy or for modifying and improving the capability of particular types of vehicle. For example, bicycles can be adapted to carry heavier loads (baskets, carriers, load frames) and the performance of traditional carts can be improved by better axle assemblies and use of pneumatic tyres.

In Bangladesh, there has been work on an improved bullock cart and harness which would allow more effective transmission of animal power and reduced injury to animals. A pedal tricycle able to carry goods, an oxtrike, has been developed by Intermediate Technology Industrial Services and its potential has been tested in a number of Developing Countries prior to initiation of local manufacture. Intermediate technology vehicles have a far greater significance in Asia than in Africa and while there is certainly scope for extension and improvement in the former, in the latter there

*Table 5.8* Low-cost vehicle performance

| | Cost index[a] | Max. load (kg) | Max speed (kph) | Max. range (km) | Route limitations |
|---|---|---|---|---|---|
| Shoulder pole | — | 35 | 5 | 20 | No restrictions |
| Shoulder frame | 10 | 50 | 5 | 20 | No restrictions |
| Wheelbarrow | 20–30 | 100–200 | 5 | 2–20 | Relatively flat |
| Handcart | 50–150 | 200–500 | 5 | 20 | Relatively flat, suitable track width |
| Bicycle | 50–90 | 40 | 20 | 60 | Relatively flat |
| Bicycle/trailer | 90–150 | 100 | 10–15 | 30–40 | Relatively flat, suitable track width |
| Pack animal | variable | 150–400 | 5 | 20 | Relatively unrestricted |
| Animal cart | 100–180 | 500–3,000 | 5 | 80–150 | Moderate hills, suitable track width |

*Source*: Based on Hathway (1985)
*Notes*: [a] indicates relative order of cost

are many areas where the introduction of such vehicles could bring about significant improvements in road transport capacity and general accessibility. In view of the cost and the lack of oil in many Developing Countries it could become essential to adopt energy-conserving technologies.

Much of the demand for transport, especially near the origins and destinations in the transport chain (field to homestead, home-stead to village), is for the movement of small quantities of varied goods over short distances – for this the simpler vehicles have distinct practical and cost advantages. However, for anything but the smallest quantities over the shortest distances the cost of transport with such vehicles will be inhibitingly expensive – and none more so than the human porter. These forms of transport therefore have the effect of reinforcing local self-sufficiency and the cost acts as a barrier to movement – there comes a time in the development process when more efficient road haulage and lower cost forms of transfer become critical.

*Plate 5.4* The Intermediate Technology Development Group has designed a variety of vehicles, such as this ox-cart in Kenya, which can be easily produced locally with available materials but greatly increase transport efficiency (Photo: Intermediate Technology/Jeremy Hartley).

## THE WIDER IMPLICATIONS

Some of the wider consequences of the development of road transport have been implied above. Thus, whilst originally vehicles may well be imported in a completed form there has been the emergence in many Developing Countries of local vehicle assembly industries. At the outset these may well assemble vehicles imported in 'knocked- down' kit form which are less costly to transport in crates or containers than finished vehicles. Often, locally constructed bodies have been added to imported chassis. With time the locally produced proportion of components increases as local industries respond to demand and eventually, as in the case of VW in Brazil, there may be a full assembly industry with a high proportion of local parts. Vehicle repair shops are a ubiquitous feature of Developing Countries – the 'shade-tree workshops' – and these are a valuable incubator of a wide range of technical skills in areas where formal education is often lacking.

The increasing demand for petroleum products, domestic oil refining, the manufacture of tyres, car batteries, bicycles and motorcycles, together with linked industries, are normal by-products of an expanding road transport industry.

Road construction and maintenance creates employment, considerable if labour-intensive methods are employed, and also a demand for equipment (which may have to be imported) or materials which may stimulate local timber, bitumen, cement or quarrying industries. The multiplier effects of road transport are potentially vast and the linkages, both forward (vehicle repairs, assembly) and backward (into agriculture – rubber for tyres), and employment generated help to give road transport its special place in the development process.

# 6

# TRANSPORT – MAKER AND BREAKER OF CITIES

Globally there is mounting concern regarding the problems of urban transport and much of the investment in transport, arguably a disproportionate share in many Developing Countries, is related to these problems. In many of the advanced economies a high proportion of the population is urban (Table 6.1) and while in many Developing Countries the proportion is much lower it has in recent decades been increasing very rapidly and in absolute terms, as in countries such as Brazil or India, amounts to a large number of people. Many of the world's largest cities are in Developing Countries and some of these have been growing at an alarming rate (Table 6.2) in relation to their resources, transport included. The number of 'millionaire' cities increased from 108 in 1972 to 204 in 1992 and is expected to be 400 in the year 2000. At the end of the Second World War none of the world's super cities (more than five million inhabitants) were in the Third World but by 1980 20 out of 30 were and by 2000 will number 41 out of 57 (Dalvi, 1986). In Nigeria, to take one example, in 1953 some 3.2 million (10.6 per cent) lived in towns but by 1990 this had increased to 40 million (39 per cent).

There is an inherent paradox in the growth of urban areas. Centralisation and agglomeration are basic features of towns but increasingly this is being eroded as urban areas spread with sub-urbanisation and out-of-town services (retail, entertainment, business) – with transport a basic causal factor in both processes. The original Central Business District (CBD), as in some American cities, may become lifeless out of office hours and it is realistic to see transport as the initial maker but increasingly, with time, the means by which the traditional city will cease to exist. Indeed,

*Table 6.1*   Urbanisation and income (1989)

| | GNP/per capita 1989 ($) | Percentage of population urban areas | |
| --- | --- | --- | --- |
| | | *1965* | *1989* |
| Low-income countries | 330 | 17 | 36 |
| Middle-income countries | 1,700 | 42 | 58 |
| High-income countries | 18,330 | 71 | 77 |

*Source*: World Bank Development Report (1991)

*Table 6.2*   Development of 'millionaire' cities

| | 1972 | | 1992 | |
| --- | --- | --- | --- | --- |
| | 1 million plus | 5 million plus | 1 million plus | 5 million plus |
| Developing Countries | | | | |
| Africa | 5 | — | 6 | 1 |
| Central and South America | 18 | 1 | 29 | 5 |
| Asia | 39 | 3 | 96 | 14 |
| Total | 62 | 4 | 131 | 20 |
| Percentage of world total | 57 | 36 | 64 | 67 |
| Other areas | | | | |
| North America | 9 | 2 | 11 | 4 |
| Europe | 21 | 2 | 24 | 2 |
| Japan | 7 | 2 | 13 | 2 |
| Australasia | — | — | 4 | — |
| Former USSR | 9 | 1 | 21 | 2 |
| Total | 46 | 7 | 73 | 10 |
| World total | 108 | 11 | 204 | 30 |

*Source*: Calculated from UN Demographic Yearbooks

Owen (1987) has suggested that if the global transport system does grind to a halt, it will be in the cities that it happens.

While accounting for only 0.5 per cent of the world's land area and in spatial terms only a small part of the global transport network, the large cities are critical to the functioning of the whole for here are the points of convergence of the national and global flows of people, goods and ideas and the location of many of the inputs – industrial, commercial, institutional, social – on which development, in the broadest sense, is based. It follows that, especially for many Developing Countries, there may be an undesirable urban bias in development strategies and this may well distort the development effort and favour 'modernisation' rather than 'distributive justice' (Mabogunje, 1980). It is difficult to see how, from a logistical point of view, it can be otherwise.

It has been argued that in Developing Countries the process of urbanisation is inextricably linked with economic growth (Dalvi, 1986) with the relationship operating at two levels. Historically, it would seem that a minimum level of urban population is linked to development as non-agricultural economic activity expands and draws people from rural areas – this Dalvi calls 'true urbanisation'. Beyond this is 'over-urbanisation' as people continue to be attracted to towns for their appearance of modernity but with inadequate associated economic growth – population rapidly outstripping employment opportunities, housing and urban services, transport among them, and sub-standard conditions for increasing numbers of urban dwellers.

Within some urban areas it would also seem to be the case that there is an undesirable bias towards addressing the transport problems rather than more pressing economic and social issues – in São Paulo when 40 per cent of the homes had no sewer connection and 60 per cent had no piped water five times more was spent by the city authority on transport than on the water-sewage system (Owen, 1987). There are probably many more cities in Developing Countries where the balance of spending is even more distorted and this is a basic problem of resource allocation which is not easily solved.

There is good reason to believe that many of the already over-sized cities will become ever larger, more crowded, less manageable and an ever-increasing drain on scarce resources simply to ensure that their arteries do not seize up completely. In a city such as Delhi, the mass of the population may never be able to afford personal motorised transport yet the streets are congested with cars and scooters. For the less affluent the transport is hopelessly inadequate and Delhi is no different from many cities in the Developing Countries. Herein lies the real transport problem.

## URBAN FORM, FUNCTION AND TRANSPORT

The demand for transport in cities, whether of people or goods, is largely determined by the spatial arrangement of the different land uses and several distinct models of urban structure have been identified (Brunn and Williams, 1983). The concentric zone model of sociologist Burgess (Figure 6.1) emphasised the core CBD and successive zones of transition (social deterioration, blight and industrial invasion), working-class housing, higher-income housing and suburbia and that of Hoyt derives from land values and rent but also has a CBD from which radiate, in relation to lines of transport, a series of wedges or sectors. The multiple-nuclei model is based on the idea that 'like' activities tend to cluster and conflicting functions repel and there will be a number of distinct focal points – this certainly seems to reflect reality in some car-oriented urban areas – Los Angeles being the best example.

These models owe much to European and particularly North American experience, where there are varying degrees of regulation, planning and land-use zoning. Cities in Developing Countries are often very different in their history. Regulation may be absent or less rigid and land use zones ill-defined and often of mixed character – there may be residential, commercial and industrial activities in the same building, not very different from the pre-industrial city identified by Sjoberg (1960). The centres of cities such as Bangkok, Beijing, Lagos and Saigon all show this diversity of land use.

As a consequence of differing histories, traditional urban structures vary greatly. Latin American cities are often characterised by a central plaza surrounded by administrative buildings, financial services and elite population. Income and quality of housing decline outwards but there may be a spine of higher income housing along favoured transport routes, often a wide boulevard, and possibly zones of 'disamenity' (Brunn and Williams, 1983). In Asia the traditional core may have been provided by a temple or bazaar but this and its associated population was often separate from a colonial walled fortress with surrounding open space. In Africa there were few areas of traditional urban culture (Egypt, south-west Nigeria) and the urban model is less clear and very varied (O'Connor, 1983) but often involving a colonial centre with adjacent but separate indigenous zones.

Rapid and sometimes massive urban growth by in-migration from rural areas is a feature of many cities in Developing Countries with newly arriving inhabitants forced to live at increasing distances from the city centre. It can be argued that in contrast with western cities the income and social class is inversely related

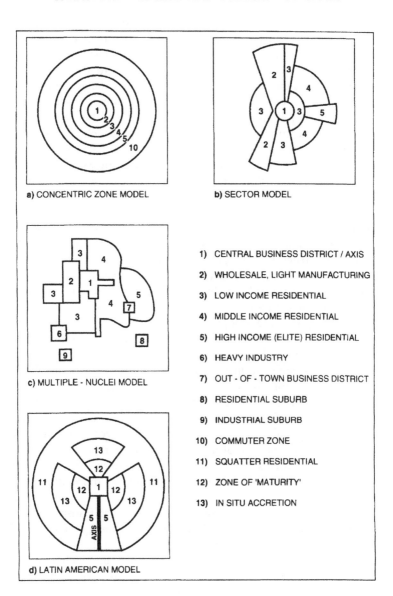

a) CONCENTRIC ZONE MODEL

b) SECTOR MODEL

c) MULTIPLE - NUCLEI MODEL

d) LATIN AMERICAN MODEL

1) CENTRAL BUSINESS DISTRICT / AXIS

2) WHOLESALE, LIGHT MANUFACTURING

3) LOW INCOME RESIDENTIAL

4) MIDDLE INCOME RESIDENTIAL

5) HIGH INCOME (ELITE) RESIDENTIAL

6) HEAVY INDUSTRY

7) OUT - OF - TOWN BUSINESS DISTRICT

8) RESIDENTIAL SUBURB

9) INDUSTRIAL SUBURB

10) COMMUTER ZONE

11) SQUATTER RESIDENTIAL

12) ZONE OF 'MATURITY'

13) IN SITU ACCRETION

*Figure 6.1*   Models of urban structure
*Source:* Modified from Brunn and Williams (1983)

to distance from the centre with high income in the core and lowest incomes on the periphery. This is of fundamental importance from the point of view of transport provision with the poorest having the longest distances to travel to the city centre (Camara and Banister, 1993). Much depends on the actual location of employment but most of the commercial, domestic service and administrative jobs are likely to be in the centre of the city although newer manufacturing industries may be more widely dispersed.

The structure of many cities in Developing Countries combines aspects of the multi-nuclei and inverse concentric models and the general lack of spatial differentiation and the overall distribution of land uses creates a dispersed pattern of demand for transport which is not necessarily satisfied by the high-capacity radial transport corridors which characterise western cities. The demand for transport, certainly in its geographical aspects, will be influenced by these city structures but also by socio-economic factors. Rapid urban growth creates an almost insatiable demand for transport especially from those least able to afford it. Whereas in advanced economies increasing income usually leads to reduced demand for public transport, in Developing Countries income growth is from a very low base, which means that it is low-cost, public transport for which demand will be increasing most rapidly.

A further significant difference is the importance in Third World cities of economic activities that may be termed informal. An International Labour Organisation study puts such informal employment at from 20 to 70 per cent of all employment (Ochia, 1990) and this is likely to produce a very different and much less regular pattern of travel demand from that which derives from the more formal array of activities in western cities. Both the spatial and temporal patterns are likely to be more diffuse in character.

Related to this connection between land use and transport is the distinction that has been made between 'walking', 'tracked' and 'rubber' cities (Schaeffer and Sclar, 1975). The walking city is the compact, pre-industrial city already mentioned, in which activities are in close proximity and easily accessible on foot or with simple hand or animal-drawn carts. Streets are often narrow and congested and, apart from a small administrative core, the functions are largely undifferentiated. Walking will be of vital importance but this may continue in the core even after the city has evolved beyond this stage.

*Plate 6.1* In many cities the movement of people and goods is mainly on foot which reduces range and capacity and may add to congestion and cost (Photo: Intermediate Technology).

The nineteenth-century urban growth of European and North American cities was invariably associated with 'tracked' transport. Railways and trams provided increases in the range and capacity of urban transport and removed the constraint which had hitherto kept residence and work place in proximity. Towns expanded spatially (Table 6.3), suburbanisation gathered pace and commuting to work became the norm. Homes were anchored to the tracks which made possible the journey to work over increasing distances and in places the railway companies were themselves involved in promoting residential property development around their stations – the Metropolitan Railway's 'Metroland' corridor north-westwards out of London provides an excellent example. Many cities developed star-like along radial rail routes – London, Baltimore and Chicago displayed this form – and functional zones became more distinct and separate.

The rubber-tyre motor vehicle brought in a third phase in the evolving relationship between land use and transport in towns. The

*Table 6.3 Zones of urban traffic influence*

| Town size (million) | Radius of zone of influence (km) |
|---|---|
| 3 plus | 56 |
| 1–3 | 48 |
| 0.5–1 | 40 |
| 0.3–0.5 | 32 |
| 0.1–0.3 | 24 |
| 0.05–0.1 | 19 |
| 0.025–0.05 | 14 |
| 0.01–0.025 | 10 |

*Source:* After Vance (1986)

car freed people from the tracks, widened the choice of residential location and led to a decline in the attraction of the traditional urban cores. The greater mobility of the work-force and the ability to service industries by more flexible road haulage likewise gave a new freedom in the location of economic activity of all types. Out-of-town shopping malls, peripheral industrial 'parks' and car commuting from a suburban or rural hinterland are all elements of this new freedom.

The transport problems of cities in Developing Countries derive in large measure from their history and in particular a rapid leap from the 'walking' to 'rubber' type often without the tracked transport which was associated with the expansion of cities in advanced economies. Bombay, in expanding northwards along the line of an inter-city rail route, is something of an exception among Third World cities in having a well-used urban rail system (Joshi, 1981). Much more typically the cities combine the incompatible characteristics of the walking and rubber phases with expansion based mainly on road transport but with road vehicles competing for space with walkers, hand carts, and animal-drawn vehicles which take up considerable space in relation to carrying capacity and determine the overall speed of movement over large areas.

While we often talk simply of 'the' urban transport problem the reality is more complex. In terms of geography and history each city is unique and so too will be the range and form of their problems. Particularly, it is unhelpful to generalise from Western cities to Third World cities where the socio-economic and historical context is so completely different and the unthinking transfer of solutions from one to the other may well cause more problems than it solves. Likewise, while one can generalise about cities in Developing Countries it is well to remember that the problems and solutions of Bombay may not be those of Mexico City or Bangkok.

Mexico City, with 15 million inhabitants, has for long had a transport policy based on urban motorway construction and road vehicles which now number 3.5 millions. Spread out in a depression surrounded by hills, the altitude of 2,250 m reduces fuel-burn efficiency and increases pollution. Yet with rapidly increasing vehicle numbers petrol consumption increased by 18 per cent between 1988 and 1992 and the city is 'choking on its own explosive growth' as it 'runs out of space to expand and air to breath' (Reid, 1992). Very different is Bombay, which is a linear city on a peninsula virtually surrounded by water with the bulk of the formal employment concentrated at the southern tip where 61 per cent of the jobs are found on 4 per cent of the area. Congestion is heavy, traffic reduced to bullock-cart speed with little road improvement and problems of maintaining even minimally acceptable standards of urban service (Patankar, 1986). Of Bangkok it has been said that the congestion is more spread out in spatial and temporal terms than would be normal in a city such as London (Cundill and Byrne, 1982) and this is certainly true for many other cities in Developing Countries. Beijing is for historical reasons distinctive in having little long-distance commuting (Lam, 1992), but it does have 7.9 million bicycles and congestion problems deriving more from non-motorised than motorised transport.

## DEMAND FOR PUBLIC TRANSPORT

It has already been noted that many cities in Developing Countries have patterns of land use which make them very different from Western cities but possibly of greater significance from the point of view of transport provision will be the socio-economic differences. While much of the wealth in Developing Countries is concentrated in the urban areas the average income levels are far lower than in advanced economies and that income is most unevenly distributed so that there are large numbers of people who cannot afford any form of transport. It has been suggested that the urban poor may be at one or other of five levels of mobility and accessibility (Eastman and Pickering, 1981):

1 The lowest rung of the urban ladder is represented by those, often newly arrived from rural areas, who cannot afford any form of transport and for whom there are no job opportunities within walking distance of the residence. Jobs rarely exist in the peripheral areas where these newcomers are forced to make their homes.
2 There is a situation in which jobs may be available but the cost of transport to get to them remains prohibitive.
3 An income level is reached at which a journey to work by cheap public transport becomes possible.
4 Income rises and expenditure on non-work journeys becomes feasible.
5 Income is such that cheaper forms of personal transport become possible – bicycle, moped – for work and other journeys.

It is generally agreed that at the lowest levels of income, all or a high proportion has to be spent on essential food, shelter and fuel. Paying for transport becomes possible only as the disposable income rises but, interestingly, that may be marginally more for those in squatter settlements, where living is very cheap, than in more formal housing (Eastman and Pickering, 1981).

For many of the inhabitants of cities in Developing Countries the ownership of personal transport is impossible and levels of access to such transport very low. Kuala Lumpur is not, relatively, a poor city but there is only one motor cycle per 15.4 persons and one car per 11.7 (Wahab, 1990). In poorer areas of the city (Eastman and Pickering, 1981) there was only one car per 26.5 inhabitants. In low-income households in Jamaica, Heraty (1980) found only one car per 49 persons and there are many cities in which even this level would be unimaginably high. In Beijing, private car ownership is almost non-existent and 45 per cent of all journeys are made by bicycle (Lam, 1992).

Access to personal transport, frequency of trips and choice of mode are all related to income levels and the range of income levels will largely determine the modal split. Eastman and Pickering's (1981) study of the poorer areas of Kuala Lumpur showed that at the lowest incomes a high proportion of the journeys were made on foot (Figure 6.2) but this decreased with rising income; over middle incomes the use of private vehicles was fairly uniform and mainly by bicycle but rose rapidly as cars came in at the highest income level. Overall, trips per day rose sharply with income with

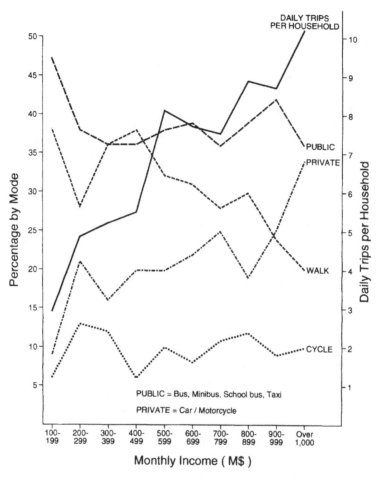

*Figure 6.2*  Modal choice and trip rates by income groups
*Source:* Based on Eastman and Pickering (1981)

the use of public transport being fairly uniform at middle incomes but high at low incomes and low at high incomes. In Accra, Ghana, it has been found (Abane, 1993) that formal sector employees on slightly higher incomes are more demanding of public transport services but that, overall, modal choice was determined more by personal factors (age, sex, income) than by characteristics of the transport (distance, time).

This pattern would seem to be representative, with slight variations, for many cities. Studies of Indian cities (Maunder *et al.*, 1981) revealed a similar income-related modal split with variations dependent largely on the extent to which public transport was available. The taxi is used mainly by the high-income group which in general does not use the cycle-rickshaw (Table 6.4). The cycle, auto-rickshaws and mini-buses are used greatly by the middle-income group while the low-income people are dependent on buses and horsedrawn tangas. Variations from this broad pattern may result from the peculiarities of particular cities – size, distances and the relative cost of different modes. Thus, in Delhi, the cycle ricksaw is restricted to the old city where there is little competition from other forms of public transport.

Modal choice is clearly related to availability but critically to cost, which per trip kilometre for Indian cities averaged 81 paise for auto-rickshaws, 34 paise for cycle-rickshaws, 17 paise for tongas, 11 paise for minibuses and 5 paise for larger buses (Maunder *et al.*, 1981) There is in this ranking a clear relationship between cost and passenger capacity and the traditional vehicles are not necessarily the cheapest. This cost factor will be the main determinant of the number of trips that are made, with low income travellers either having to take the cheaper modes or walk. Thus in squatter areas of Delhi not well served with public transport, 58 per cent of all journeys were on foot whereas in a higher-income, more spaciously planned development, with reasonable public transport, the proportion of walking journeys dropped to 26 per cent (Maunder, 1983).

*Table 6.4*  Monthly income and modal use (percentage of users)

|  | Monthly income (Rs) | | |
| --- | --- | --- | --- |
|  | *Low* <500 | *Medium* 501–1,000 | *High* >1,000 |
| Cycle rickshaw | 35 | 55 | 12 |
| Auto-rickshaw | 17 | 55 | 31 |
| Tonga[a] | 56 | 35 | 12 |
| Phut-phut (Delhi) | 22 | 40 | 38 |
| Minibus | 39 | 47 | 15 |
| Bus | 62 | 30 | 9 |
| Taxi | 10 | 12 | 78 |

*Source:* After Maunder *et al.*, (1981)

*Note:* (a) Horse-drawn or motorised in several cities

Studies in the Philippines and Indonesia demonstrate the same positive relationship between expenditure on transport and household income (Ocampo, 1982) but what must be emphasised is the large number of people who are forced to live in inaccessible, peripheral city areas which are poorly served by public transport and whose incomes restrict their modal choice and therefore mobility. Further, this already large number is daily swollen by in-migrants: an 8–9 per cent annual city population growth rate is not unusual and 5–6 per cent is quite normal.

## BUS TRANSPORT

From the above discussion it emerges that the bus is likely to be the appropriate urban transport to satisfy a large low-income demand. It is worth considering this further.

The growth of few cities in Developing Countries was based on high capacity urban rail systems and the bus is therefore the essential element in arterial and feeder services. Indeed, buses have an operational flexibility that allows them to satisfy rapidly expanding, spatially dispersed, low-income markets, where demand is increasing faster than can be satisfied by growth of private vehicle ownership. It has been estimated that a 1,000 increase in population creates a demand for 350 to 400 trips per day and every square kilometre of urban expansion an extra 500 daily trips (Jacobs, et al., 1986). While some cities in Developing Countries are developing mass transit rail systems (to be described below) road based solutions will of necessity dominate for a long time to come and will be required to satisfy rapid growth in demand, whereas declining bus patronage has been the main problem facing bus operators in advanced economies.

For a sample period 1974 to 1983 (Table 6.5) there was substantial expansion of bus fleets, routes operated and passenger trips for a selection of cities in Developing Countries in contrast to contraction in Britain. Yet in most Developing Countries service provision is still well below that of the advanced economies (Table 6.6) – in India's cities it was less than half that of UK cities (Fouracre et al., 1981). As always, these averages conceal great variations and in Indian cities the number of buses per 100,000 population ranged from 3.3 (Kanpur) to 42.6 (Trivandrun) and would appear to have been inversely related to the availability of 'intermediate' forms of public transport.

*Table 6.5*  Bus use changes in selected cities, 1976–1983 (percentage change over period)

|  | *Bus fleet* | *Routes operated* | *Passenger trips* |
|---|---|---|---|
| British Isles | −2.2 | — | −3.2 |
| Addis Ababa | 0.6 | 2.0 | 7.4 |
| Bangkok | 1.5 | 5.9 | 16.2 |
| Bombay | 7.8 | 7.1 | 9.0 |
| Cairo | 8.8 | 4.2 | 1.0 |
| Delhi | 24.9 | 18.3 | 24.3 |
| Madras | 25.1 | 21.8 | 62.0 |
| Nairobi | 9.5 | 4.0 | 8.3 |

*Source:* Jacobs *et al.* (1986)

In many cities in Developing Countries buses provide the main means of transport for a substantial proportion of the journeys. In Kingston, Jamaica, it has been estimated that 43 per cent of all work trips (Heraty, 1980), in Kuala Lumpur 34 per cent (Eastman and Pickering, 1981) and Indian cities an average 38 per cent (Fouracre *et al.*, 1981) were by bus, with dependence on this mode being even higher for the lower income groups although, overall, people in this group made less journeys.

What is under consideration here are the larger types of single or double-decker buses with passenger capacities in the range of 70 to 100 or possibly as high as 150 in articulated versions. In response to rapidly growing demand vehicles with longer wheel bases, higher loading weights and in places wider bodies have been introduced. Articulated buses require more skilled drivers and specialised maintenance and in the Developing Countries the heavy over-crowding and usual adoption of labour-intensive fare collection

*Table 6.6*  Bus provision – 1980s

| Buses per 100,000 population | |
|---|---|
| Africa | 30 |
| India | 30 |
| Asia | 48 |
| Other LDCs | 63 |
| UK | 90 |
| Bus route km per 100,000 population | |
| Developed (average 94 cities) | 100 |
| Developing (average 54 cities) | 54 |

*Source:* After Jacobs *et al.*, (1986)

can make for problems (Crawford, 1990). It is also worth noting that there is not always a clear distinction between passenger and goods transport and provision for considerable quantities of the latter on passenger vehicles may be an essential requirement.

In 1989 in Kinshasa, Zaire, an articulated bus was introduced with a nominal capacity of 250 passengers and a manufacturer's claim for a record 440 (*West Africa*, 22/1/1990)! These five-axle vehicles are a combination of 14-m York trailers modified for passengers and DAF tractors assembled locally and provide low fares and continuous music! These were introduced by a private company following the collapse of the state-owned bus company and at a time of considerable urban unrest thought to be prompted, at least in part, by dissatisfaction with serious problems of access from increasingly distant suburbs. The mainly flat urban terrain and wide boulevards favoured these larger vehicles and their success in Kinshasa led to their introduction in Lubumbashi. More widespread adoption of this technology depends on a suitable pattern of urban highways – something many cities will lack.

In the operation of bus fleets in towns of Developing Countries several factors need to be taken into account. Labour costs are relatively low and labour-intensive methods can be adopted; fuel costs are usually a high proportion of total costs and must be reduced to a minimum; and capital and maintenance costs are high (Barrett, 1988). Standardisation makes for continuity of experience, more effective use and lower overall cost of spares and maintenance facilities but adopted too rigidly may inhibit desirable innovation and the ability to service all types of route. The intensity of operations, high loadings, congested road space and environmental factors (temperatures, rainfall, dust, road damage) all increase operational and maintenance costs in situations where foreign exchange may be in short supply and lead times for obtaining spares and equipment often to be measured in months, even years, rather than days or weeks. Many vehicles may be out of operation for long periods of time.

Revenue is limited by the low-income market served and often, as in Karachi (Faulks, 1990), by government imposed ceilings on fares for social reasons. There is also evidence of considerable 'leakage' in revenue collection systems (Badejo, 1990) through staff fraud, passenger evasion and physical problems of collection. One study found an unbelievable 12 persons per square metre on Bangkok buses (Crawford, 1990) and it is not unusual to have more than one fare collector per vehicle (Barrett, 1986) and still have problems.

Traditionally, urban bus services have been provided by public agencies or large private companies but there is certainly a view that they are provided more effectively by smaller operators or even owner-drivers. However, such operators are likely to use smaller buses and can less readily accommodate very heavy demand. They are also likely to charge higher fares, while maintenance and safety standards may be compromised and larger numbers of vehicles added to already congested roads. In Karachi, private-sector buses have been shown to operate irregularly, sometimes illegally and often dangerously to cut costs and increase revenue (Faulks, 1990) and while in theory competition may stimulate efficiency, unregulated small-scale operations may reduce efficiency and also cut the revenue for the larger operators.

For planning purposes and on the basis of 75-person units it has been calculated that one bus is needed for every 1,500 to 2,000 persons (Faulks, 1990) but while this figure may be an ideal it is rarely attained in most Developing Countries. In a selection of Indian cities in the late 1970s, but probably not very different now, the average was 0.4 conventional buses per 2,000 population (Fouracre *et al.*, 1981) and demonstrably this will be grossly inadequate for the demand that exists in such cities and of necessity a variety of 'intermediate' public transport vehicles fill the gap in conventional bus provision.

## INTERMEDIATE PUBLIC TRANSPORT (IPT)

In recent years a range of terminology has been introduced for what is here called 'intermediate public transport' – 'informal', 'unincorporated', 'para-transit' and 'low-cost' (Ocampo, 1982) to which has now been added 'informal high-occupancy' (Wright, 1992). All are used more-or-less interchangeably for a variety of vehicles and operational systems which fill the gap between the mass-transit systems, whether road or rail, and the non-motorised transport.

Not only is there a range of terminology, but there is also a great variety of technology (Table 6.7) and many cities in Developing Countries have developed their own, often highly distinctive, forms of intermediate public transport. It has been said of the

Table 6.7   Selected types of Intermediate Public Transport

| | City | Vehicle type | Local name | Passengers |
|---|---|---|---|---|
| Côte d'Ivoire | Abidjan | Minibus | Gbakas | Variable |
| Ghana | Accra | Minibus | Tro-tro | Variable |
| Kenya | Nairobi | Minibus or pick-up | Matutu | Variable |
| Sudan | Khartoum | Minibus | Bakassi | Variable |
| Zaire | Kinshasa | Minibus, pick-up | Kimalu-malu | Variable |
| Zimbabwe | Harare | Large estate | Emergency taxi | 4–8 |
| Hong Kong | Hong Kong | Mini-midi buses | Public Light | 16–24 |
| India | Various | Minibus, auto-rickshaws | | 16+, 2–6 |
| Indonesia | Jakarta | Minibus (fixed route) | Opelet | 10–17 |
| | | van (fixed route) | Bemo | 7–15 |
| Jordan | Amman | Larger cars | Servis | 6–10 |
| Malaysia | Kuala Lumpur | Minibus | Bas mini | 20–35 |
| Philippines | Manila | Re-modelled jeep | Jeepney | 10–14 |
| Thailand | Bangkok, Chiang Mai | Converted pick-up | Silor | 10–12 |
| | | 3-wheelers | Samlor | 2 |
| Turkey | Ankara, | Larger saloon | Dolmus | 7–10 |
| | Istanbul | Minibus | Minibus | 8–14 |
| Bolivia | La Paz | Midibus | Microbus | |
| Puerto Rico | San Juan | Minibus | Publicos | |
| Venezuela | Caracas | Midibus | Porpuestos | |
| Jamaica | Kingston | Minibus | Minibus | |

Note: For all these vehicles there are often considerable differences between the legal and actual carrying capacities.

jeepneys of the Philippines that they are the 'most visible form of popular art, embellished by gaudy decorations, mirrors and inscriptions' (Wright, 1992) and the tro-tro of Ghana, while rarely so elaborately decorated, invariably has its slogans.

For the most part IPT would appear to have emerged spontaneously to fill perceived gaps in the conventional transport provision in urban areas (Figure 6.3). Such gaps have become more obvious as the urban areas and populations have expanded and the existing public transport has proved quite inadequate to satisfy the

*Plate 6.2* The inadequacy of conventional bus services in many cities, here in Bangkok, creates a demand for Intermediate Public Transport of various types.

new levels of demand. Increased crowding, inadequate investment, insufficient maintenance and in some cases poor management have combined to reduce efficiency and capacity. Service provision has also declined and there has been a growing demand for alternatives – Cairo, Caracas, Kingston, Lagos and Lusaka are a few of the many cities for which this has been documented (Coombe and Mellor, 1986).

The inability of conventional bus services to satisfy growing demand, while undoubtedly important in many cities, was not the only factor encouraging the expansion of IPT. Thus, privately owned mini-buses were introduced into Kuala Lumpur in the mid-1970s although the city was served relatively well by frequent, low-cost, stage-bus services supported by cheap taxis. Demand for licences far in excess of the number issued (Walters, 1979) suggested that this was seen as a possibly profitable investment by local entrepreneurs, which created jobs and the demand for 'shade-tree' servicing facilities. Being non-unionised the mini-buses had a flexibility lacking in municipal operations and they proved

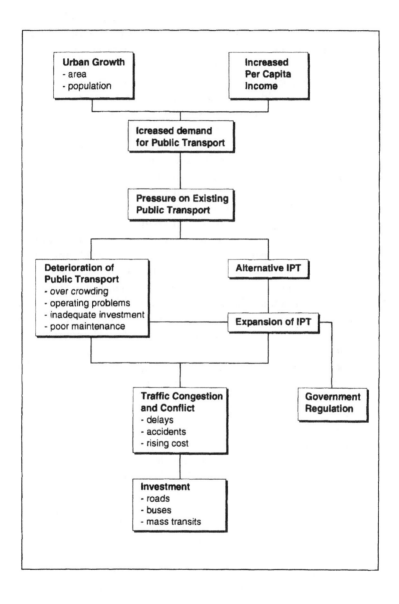

*Figure 6.3* Development of urban transport systems

invaluable on lightly used routes and as short-haul feeders to main bus services, from some of which routes they were excluded. Officially, the mini-buses were seen as a way of persuading those with cars to leave them at home – with what degree of success is unclear – but by the late 1970s accounted for 30 per cent of all trips.

Coombe and Mellor (1986) suggest that the wide range of income between car owners and the poorer sections of the population in Third World cities calls for a wide range of transport services. A bus sufficiently cheap for the poor will provide a crowded, uncomfortable service and is unlikely to be attractive to middle-income travellers without cars and will certainly not entice people away from their cars. Further, the nature of the roads in congested inner-city areas and the dispersal of population and economic activities on the urban fringes are not easily served by larger buses operating fixed routes. It is a virtue of the IPT services that they offer flexibility in spatial and temporal terms (Table 6.8) which is lacking in mass-transit systems. The degree of flexibility clearly depends upon the extent to which the urban authority attempts to control and regulate the IPT services.

There are of course many Third World cities in which there is little or no competition between conventional bus operations and IPT for the simple reason that the former is largely absent. Thus, only a few Nigerian cities – Lagos, Ibadan, Port Harcourt, Benin and Kaduna – have what might be termed conventional, municipal bus operations and in many Indian cities, some of considerable size, such services play an insignificant role (Fouracre *et al.*, 1981). In all such cities, IPT provides all or the bulk of the public transport.

In view of the spontaneous emergence of IPT in response to particular urban conditions it is not surprising that the vehicles and operational characteristics are extremely varied but certain common characteristics can be identified (Coombe and Mellor, 1986) and these are summarised in Table 6.8.

The great variety of vehicles has already been noted. Some, such as the rickshaws of Indian cities and the becaks of Indonesia are human-powered and will be considered below but for the most part we are concerned here with motorised vehicles of small or medium size (Table 6.7). At their simplest, these may be a small-engined tricycle, such as the samlor of Thailand, the auto-rickshaw common in many Indian cities and the tuk-tuk of Indonesia.

Vehicles such as the dolmus of Turkey are essentially saloon cars

*Table 6.8*  Operational characteristics of conventional buses and
minibuses

| Conventional buses | Minibuses |
|---|---|
| Large, monolithic, inflexible, bureaucratic, corporate organisation (state, municipal, private) | Small, flexible, minimum administration (owner operator, small private firms, some municipal) |
| Highly regulated | Often limited or non-enforced regulation |
| Lack of competition, often unprofitable | Fierce competition, mainly profitable |
| Large scale fare avoidance, fraud | Minimal fare avoidance |
| Larger, relatively uniform vehicle types, restricted access, fixed routes, spatial inflexibility | Smaller vehicles of great variety, wider access, fixed and variable routes, spatial flexibility |
| Longer stops to fill capacity | Fewer, shorter stops |
| Slower service | Faster service |
| Less frequent service | More frequent service |
| Longer passenger wait | Shorter passenger wait |
| Temporal inflexibility | Temporal flexibility |
| Slow response to changing demand | Rapid response to changing demand |
| Better maintenance, better safety record | Poorer maintenance, bad safety record, reckless driving |
| Adversely affected by minibus competition and congestion adds to costs | Can respond to competition from conventional buses |
| Higher element of imported technology, capital, skills, management | Scope for local enterprise, capital, management, skills |

*Source*: After Rimmer (1984); Lam (1992)

with little modification while the silors of Thailand are pick-up
trucks converted for passenger use, with bench seats, a metal or
canvas roof and a tail gate replaced with a passenger boarding step.
The Philippine jeepney may have had its origin in and taken its
name from surplus Second World War jeeps but conversion locally
is now considerable with local components, assembly and body-
building shops making significant inputs – one establishment,

Sarao Motors, makes eight vehicles a day, employs 380 workers and produces almost everything except the engine (Ocampo, 1982). The matutu of Kenya may be a modified pick-up and the molue of Nigeria is usually a locally converted, small lorry.

Of growing importance have been a new range of off-the-peg mini-buses, mainly imported, such as VW, Ford, Benz, and with little modification apart sometimes from decoration and with seating capacities of 10 to 25 persons. These are now found widely and have become the main form of IPT in Kuala Lumpur and Hong Kong (the Public Light Bus). The gbakas of Côte d'Ivoire, the tro-tro of Ghana, the bakassi of Sudan and the publicos of Puerto Rico are all of this type and in general they provide more comfortable seating than some conversions but overcrowding can soon reduce this advantage.

Because of their small size, these vehicles are more quickly filled with reduced stopping time and passenger waiting and they often provide very frequent services and temporal flexibility (Figure 6.4). The size is also an advantage in negotiating narrow, congested streets and while some, especially where there is greater regulation, may operate fixed routes, it is often the case that deviation is the norm (Lam, 1992). Many LPT vehicles, especially of the smaller kind, operate on a 'shared-taxi' basis, in which routes are determined purely by demand.

Similarly, the size of the vehicle and the possibility of starting with one second-hand reduces the capital investment and places it within the range of local, private initiative. Coombe and Mellor (1986) found that, typically, 90 per cent of the vehicles were owner driven or 'hired' from family members, although there are certainly places where multiple-vehicle ownership and operation or leasing has developed or where the individual owners group into cooperatives. Nevertheless, entry to the market is usually relatively unrestricted except where the authorties impose restrictions on numbers – with a little enterprise even that may not be an insuperable barrier!

The owner-operator and small firm therefore characterise the IPT sector, competition can be fierce, administrative overheads are minimised and there is rapid response to shifts in demand. The services are unlikely to be particularly profitable and they will only stay in operation as long as they are, at least, marginally so. However, they are unlikely to be a financial burden on the community at large, like many of the larger, conventional bus opera-

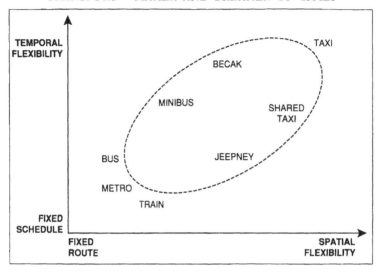

*Figure 6.4* Spatial and temporal flexibility of transport options
*Source*: After Lam (1992)

tions. There is little scope for cross-subsidisation and services will be reduced at times of low demand. The small-vehicle intimacy of driver and passenger reduces the fare avoidance and fraud which seem to typify many larger public transport operations. It may be necessary for drivers to work undesirably long hours to cover their costs and make even a meagre living and the IPT sector has been criticised for severe over-loading of vehicles, dangerous driving habits (poor lane discipline, double parking) and the lack of proper maintenance – all of which contributes to a poor safety record.

## REGULATING IPT SERVICES

It is often the case that initially the IPT sector operates illegally, being neither properly licensed nor insured for the services they provide. Only when they begin to provide a source of revenue to municipal authorities is some sort of licensing eventually imposed. What are now the regulated Public Light Buses of Hong Kong originally started in the 1960s in the New Territories, where a variety of dual-purpose vans, passenger lorries and private cars operated illegally as public service vehicles to provide capacity that was so patently lacking in the official public transport system (Lee, 1990).

Although individually small units, by sheer force of numbers the IPT services are often a major contributory factor to urban road congestion. A strike of mini-bus drivers in Hong Kong in 1974 allowed the conventional bus and tram operators to achieve their scheduled mileages – a rare feature of their operations (Coombe and Mellor, 1986). Not surprisingly there has in many places been a demand for a greater measure of control over IPT operations which may be viewed as a cause rather than the solution of urban transport problems.

In Hong Kong the IPT services, while an increasingly necessary element in total transport provision, were technically illegal until 1969, when regulatory measures were introduced to protect passengers, ease congestion and provide a better coordination between the different sectors of public transport (Lee, 1990). The total number of PLBs was restricted, and the dimensions, weights, passenger capacities and liveries became controlled, as were boarding and alighting arrangements. Routes are franchised and there is continuing negotiation with operators on the introduction of new services. The whole may be seen as a pragmatic approach to planning, with the government monitoring, regulating and controlling but encouraging private investment.

In the early 1980s, Hong Kong investors introduced mini-buses into the Shenzhou Open Economic Zone and their use spread to other cities. In Beijing, where there is little private capital available, the five main mini-bus companies are in effect government-owned. They operate as official alternatives to conventional buses although there is evidence of some flexibility in routeing and even of some non-official operators (Lam, 1992).

In Harare, 'pirate' taxis were legalised in 1980 and became 'emergency' taxis, ostensibly to cope with peak-hour deficits in public transport supply and while the name persisted for some time the vehicles now have to be properly registered, the number per route is in theory controlled, routes are marked on vehicles and insurance, roadworthiness and vehicle capacities are regulated (Le Fevre, 1981). In Santiago, Chile, a rapid increase in the number of mini-buses and taxis after 1978 was the result of some liberalisation but operators have to be approved and licensed and numbers are limited. The main aim of the city authority was the elimination of 'socially negative practices' (Fernandez and De Cea, 1991).

In Bangkok, registration is necessary, with an annual fee, but this provides only a limited control over the supply in different parts of the city and there is a tendency for the vehicles to concentrate on the downtown, high-demand areas rather than peripheral areas where the need may in fact be greater. In Manila, the jeepneys have been encouraged to join in cooperatives but an official total of 28,000 jeepneys is thought in practice to be less than half of the real number. Jakarta has attempted to phase out the 'shared-taxi' and consolidate the mini-buses and taxis into cooperatives – a practice that has spread to other Indonesian cities. Even where not imposed or officially encouraged it is often the case that the IPT operators band together in a spontaneous way to protect their interests – this may be on the basis of a route, group of routes, or, in the case of a small town, of all the operators.

It seems that even where there is a measure of regulation, much of the IPT operates either on the margins of, or, even outside, the regulatory codes, which are strictly imposed and comprehensively enforced in but few places. Indeed, it could be argued that it is the

*Plate 6.3* Intermediate Transport becomes an essential element in urban transport but in Bangladesh, as in many other places, it also provides a vehicle for local artistic enterprise (Photo: Intermediate Technology).

informality of the IPT system and its ability to operate outside established regulatory framworks which is both its main advantage in providing transport services geared to changing local demand but also its main disadvantage when it comes to the wide control and management of urban road transport in congested cities.

## THE 'FORMAL'/'INFORMAL' DEBATE

There has been a growing concern at what in places are seen to be attempts to squeeze out the IPT operations. Rimmer (1984) pointed to the apparent paradox that, while IPT-type services were being encouraged in some advanced economies (taxis, community buses, dial-a-ride, car pools), as a solution to some of the problems of transport provision, there were Developing Countries in which they were being discouraged. If development is seen as a process whereby less modern must be replaced by more modern then it is possible to view IPT as undesirable in its lack of incorporation, possibly in being exploitive or even humiliating in its use of people – most obviously so in the case of non-motorised forms (pedicabs etc). A modernisation approach to development results in an emphasis in large-scale, incorporated, highly capitalised, 'western' style transit systems, with high fixed costs, often considerable subsidy and great dependence on imported equipment, skills and capital – what has been called 'the imperialism of public transport' (Dick, 1981).

Post-modernists would argue that in the context of the Developing Countries IPT has characteristics and qualities that, if anything, should be encouraged rather than extinguished, providing as it does a technology, which in the best sense, is appropriate. IPT provides valuable employment, in places for many thousands, and therefore capitalises on a resource that is plentiful and it economises on resources that are scarce – capital, technology and managerial skills that may well have to be imported at great cost. IPT provides a varied, multi-level system of operations, vehicles and services where in contrast the conventional mass transit emphasises monolithic uniformity and bureaucracy (Figure 6.5). It is possible for IPT to provide users with services appropriate to their demand. Can Developing Countries afford to subsidise, as invariably seems necessary, their conventional transport operations?

The changes involved in the move from IPT to conventional

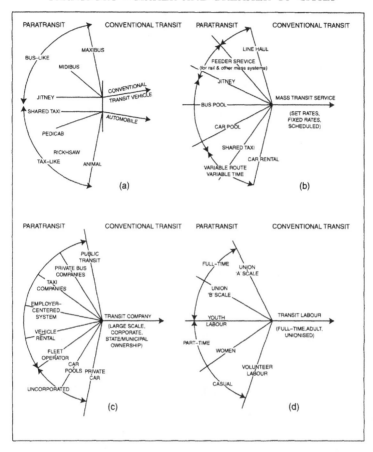

*Figure 6.5* Paratransit and conventional transport – operating contrasts
*Source*: Based on Britton (1980)
*Notes*: (a) vehicles (b) services (c) operators and providers (d) labour

mass transit have been represented by Britton (1980) and in modified form can be seen in Figure 6.5. In terms of frequency of service, reduced waiting time and speed, IPT may provide superior service at the same fare (Walters, 1980).

As in many conflict situations the answer is probably somewhere between the two extremes, in this case with conventional and IPT operating in a coordinated and complementary manner. There are situations in which IPT can provide specialised line-haul services (contract services for workers, additional peak hour capacity, services to supplement decline in off-peak provision by conventional transit) and others in which it could operate in extension of conventional services (shorter-distance services in low-density areas, feeders to high-density routes) with recognised interchange facilities to emphasise the associated roles. IPT services, as in Kuala Lumpur, are used to provide quality service alongside conventional transit, it is not necessarily a second-class service, and there are places like Chiang Mai and some Indian cities, where for lack of conventional transit, the IPT provides an area-wide transit system.

It has been argued that viewing IPT in this way takes one away from dualistic modernisation versus extreme post-modernism approaches and instead emphasises the possible interrelationship between the modes in terms of the mobility provided, the accessibility requirements and wider community issues (Rimmer, 1984). This idea has been developed more recently by Wright (1992) when he argues that urban transport makes sense only when it is considered as an organic whole in which successful solutions will combine key elements of three sets of characteristics – those relating to the modes of transport, those relating to the city and those relating to the people. This will be discussed at greater length below.

## NON-MOTORISED TRANSPORT

Even in those countries with sophisticated levels of transport provision it is invariably the case that the shortest distances are covered on foot, a fact not always appreciated in transport planning. As already noted, the proportion of walking journeys and transport by non-motorised forms of transport usually increases in an inverse relationship with income and in many Developing Countries non-motorised travel for lower income groups is overwhelmingly important.

At the simplest level this is just walking or riding a cycle but there is a great variety of human-powered utility vehicles (HPUVs) for passengers and freight often operating as public transport – rickshaws, hand carts, cycle-based rickshaws and cycles with side-cars (Hathway, 1985). There is also the use of animals for direct carriage of goods and people and the use of animal-drawn vehicles with an almost limitless variety of local styles. Here is the transport technology reflecting local environmental and resource conditions at the most obvious level. These non-motorised forms of transport are a conspicuous feature of the urban areas and have a significance which is out of all proportion to their level of technological sophistication. Calcutta has an estimated 30,000 rickshaw pullers and a study of Phnom Penh, where the average annual per capita income (1983) was $60, showed non-motorised transport was of critical importance, with around 10,000 'cyclos' (cycle rickshaws) being a main element. The same study (Etherington, 1993) suggests that these cyclos are under increasing pressure from motorcycle based vehicles – a first stage in the modernisation process.

Rimmer suggests (1982) that the smallest town in which an intra-urban public transport system could have a separate existence would have a population of 10,000 – for all smaller towns transport is likely to be a by-product of rural–urban and inter-urban links. For all such smaller settlements – and indeed in many larger ones where public transport is inadequately developed – the non-motorised forms will play a significant role possibly alongside IPT. The Peoples' Republic of China provides a good example. Here bicycles are the main form of personal transport. In 1988 there were 220 million bicycle owners in China and in the 23 cities with over one million inhabitants there were 80 million bicycles, with over five million in Beijing alone (Zhihao, 1990). It is estimated that with 120 bicycle factories producing 20–40 million units a year, the number in use could rise to 500 million by the year 2000.

A 1983 survey of factory workers in Beijing found that 37 per cent of work journeys were made on foot and 30 per cent by bicycle. This modal choice was influenced by a pre–1980 prohibition on private car ownership and by a workers' travel allowance, which could be accumulated for the purchase of bicycles (Zhihao, 1990). The relatively flat terrain of a number of the larger cities (Beijing, Shanghai, Tianjin) undoubtedly favours cycling, as does the serious inadequacy of public transport in the face of rapidly rising demand and urban roads which cannot easily accommodate larger vehicles – in Beijing an estimated 12 per cent of the urban road network cannot take buses.

In many Indian cities there are also very high levels of cycle ownership and a survey in medium-sized cities (Fouracre and Maunder, 1986) showed that 60 per cent of the households had access to at least one bicycle, with Delhi now having over one million on its streets and the number growing at 4 per cent a year.

In many Indian cities bicycles account for 35–50 per cent of the traffic on main corridors, with one junction in Delhi having a flow of over 7,500 bicycles an hour. Not without reason, 'cycle cities' has been used to describe cities such as Pune, Kanpur and Lucknow.

Research has shown that in India bicycle ownership is in fact spread fairly evenly between the income groups but in terms of actual use the lower-income groups dominate and especially for journeys to work – in Delhi for the lowest-income group, 40 per cent of the work journeys in the 2–8 km range are by bicycle, in contrast with only 10 per cent for the higher-income group (Maunder and Fouracre, 1989). As cities become larger and distances and journey longer, the bicycle becomes less attractive.

While income levels in many African urban areas are comparable with those of many Indian cities, bicycle ownership and use is generally less significant, with only 6–15 per cent of households having access to a bicycle. It would seem that there are considerable variations in the way in which bicycles are perceived but why this should be so is not easily explained. Whereas China and India have very large domestic bicycle manufacturing industries only a few African countries have even small assembly plant and the cost of imported machines is high – this reflects the generally small population and, therefore, small market in most African countries. Maunder and Fouracre (1989) suggest that whereas in India bicycle riding would appear to have status value, in Africa it is considered demeaning and possibly more dangerous – observation lends subjective support to this idea. However, observation also suggests that even within Africa there are regional variations in bicycle use – in eastern Nigeria bicycles have always been more popular than in western parts of the country and in francophone more than anglophone West Africa, with the implication that there are cultural factors operating.

The pedal bicycle may also be used for 'public' transport of people or goods and there are many local variants of the bicycle with side-car and the tricycle (Hathway, 1985) – the 'cyclos' of Phnom Penh and the 'becaks' of Indonesia have the passenger seat in front of the pedaller, while in the trishaws of Indian and Bangladeshi towns it is usually behind. Such tricycles are able to take one or two, and exceptionally, three passengers. Rimmer (1982), in his study of 'pedicabs' in Malaysia, found the highest concentrations in states with the lowest GDP, higher-income settlements having the lowest proportion with the exception of Georgetown where, arguably, a tourist demand kept the pedicabs in business in a relatively high-income area.

Studies in India (Fouracre *et al.*, 1981) show that average journeys by cycle rickshaws tend to be over distances of less than two kilometres but with high load factors, journeys often being refused if there is no chance of a return fare. The fares tend to vary greatly between towns but on average are lower than for motorised IPT but more than for buses. In general this is the poor persons' transport.

A number of positive reasons can be advanced for encouraging the use of bicycles as a form of private or public transport (Zhihao, 1990). Over short distances in the rush hour bicycles often provide the quickest form of transport and require relatively little parking space. Their use conserves energy resources and does not add to air pollution. Repairs are easily effected and the bicycle is suitable for a wide age range and for many who could not drive a motor vehicle. Bicycles clearly provide a flexible, demand-responsive mode which is cheap to operate and makes little demand on scarce resources. Wright (1992) in his assessment of the qualitative performance of the different modes gives cycling more 'superior' grades than any other form of transport (see Table 6.10).

A number of counter-arguments can also be advanced. In particular, the slow, non-mechanised transport is seen to be a major contributor to urban road congestion and accidents – in China 60 per cent of all road accidents involve bicycles but this is clearly related to the level of use and is likely to be exceptional for Developing Countries overall. In Indian cities the private car is not a significant contributor to road congestion and of far greater importance is the disorganised juxtaposition of pedestrians, bicycles, hand carts, animal-drawn carts, rickshaws, buses and taxis. This has also been noted in African cities (Abane, 1993). This is possibly an argument for the more effective segregation of different types of traffic rather than the elimination of the bicycle (as has happened in Maharashtra State in India, some Indonesian cities and in parts of Phnom Penh) or any other particular type of vehicle. Some have argued that the pedicab pedallers represent a disadvantaged low-income group with few prospects and are parasitic upon the urban, modern sector while the users are subject to unsatisfactory travelling conditions. Modernisation, it is argued, therefore requires the removal of this undesirable informal sector.

Others argue (Rimmer, 1982) that this informal sector activity is vital and dynamic and capable of providing valuable employment for increasing numbers of rural–urban migrants – 80 per cent of Phnom Penh's riders were rural migrants and 91 per cent of these rented their bicycles as a way of establishing themselves in the urban area (Etherington, 1993) and remitted the bulk of their income to the rural home. However, Etherington also noted that the cyclos were under threat from motorised transport and the government tended not to be sympathetic – this is certainly the case also in many other Developing Countries. It would certainly be unwise to force the elimination of the non-motorised modes until such time as more sophisicated and higher capacity transport can be provided at a price that is affordable – economically, socially and one must now add, environmentally. Ironically, in many advanced economies there is mounting pressure for the restructuring of urban networks to accommodate the environmentally friendly bicycle!

## RAIL-BASED TRANSIT SYSTEMS

We have already noted that in contrast with industrialised countries the growth of few cities in Developing Countries was based on tracked transport. Nevertheless, some did have trams (Calcutta, for example) and many more now see the provision of rail mass transit as the solution to their urban transport problems. This can be seen as the inevitable march of modernisation but may also be seen as a gross mis-allocation of resources and technology in the context of most Developing Countries. The situation has been admirably put in context by Ridley:

> Metropolitan railways are particularly expensive, Developing Countries are particularly poor . . . anyone engaged in metro design would do well to understand the political and social environment in which decisions are made
>
> (Ridley, 1986)

The characteristics and also some of the inadequacies of existing transport systems in many rapidly growing Third World cities will be apparent from the foregoing discussion. Many are dependent on a combination of buses, IPT and HPUVs usually operating in a largely unplanned, uncoordinated and inadequately regulated manner and there is deficiency in financial, human and infrastructural resources to provide and maintain satisfactory levels of public transport provision. The problems become worse by the day as overcrowding mounts and infrastructure deteriorates. In the face of these apparently insuperable problems it is not perhaps surprising that some transport planners start thinking in terms of a technological 'big push' to break into the spiral of decline. Rail-based transit systems are seen to provide just this.

The London 'underground', by no means all of which is underground, provides an early model but world-wide there are now over 100 'metros' in operation or under construction, many of them in cities in Developing Countries. Rail-based mass-transit systems come in many forms and range from the tram which shares road space with other traffic (as in Calcutta), through partially segregated track at grade to exclusive, dedicated track which may be elevated or in tunnels and often combines both (e.g. Hong Kong, Singapore). Likewise, the vehicles themselves vary from 'light', able to negotiate sharp bends and steep gradients (e.g. London's Dockland Light Railway) to conventional 'heavy' rail technology (Table 6.9). Given that many cities in Developing Countries do not have extensive urban rail networks it is likely

*Plate 6.4*  The tram, as in Calcutta, may be a more appropriate and a less costly form of rail transit than metro systems.

229

Table 6.9 Rail-based transit systems

| | Tramway | Light rail | Rapid rail | Suburban rail |
|---|---|---|---|---|
| Technology | Fixed rail, tram, streetcar, single/double units, mixed traffic on city streets | Light rail units, articulated/trains, fixed rail on street, mixed or segregated | Light rail units, trains, segregated track (surface, elevated, underground) | Conventional rail or metro-type stock, possibly track shared with inter-city trains |
| Speed/capacity | Slow (<12 kph) | Up to 25kph | Journey average 30–35 kph | Journey average 45–55 kph |
| | 100–200 pass/unit 6,000 pass/hour with single unit mixed traffic 12,000–15,000 on exclusive track | 800–900 pass/unit 20,000 pass/hour street level, 36,000 pass/hour on segregated track | 1,500–3,000 pass. 70,000 pass/hour per line | Shared track – 10,000–20,000 pass/hour one direction, 48,000–72,000 on exclusive track |

| | | | | |
|---|---|---|---|---|
| Capital | Track/power – $4 million/km Trains $300,000 | Track/power/synch. – $6–10 million/km | Elevated track – $23–50 million/km, tunnel track – $63–98 million/km | Track $6–10 million Trains $6 million |
| Operating cost (US cents) | 2–8 pass/km | 8–10 pass/km | 10–15pass/km | 5–10 pass/km |
| Total cost | 3–10 pass/km | 10–15 pass/km | Surface 10–15 Elevated 12–20 Tunnel 15–25 | 8–15 pass/km |
| Revenue | Fare box cover of costs unless revenue 'leakage' | Usually cover operating and sometimes total costs | Farebox revenue may cover operating but rarely total costs | Some cover operating costs, few cover total cost |

*Source:* Modified from World Bank (1986)

that new developments will be at the lighter end of the scale. However, even these may vary considerably in design, capacity and operations and a number of writers (Catling, 1986; Bonz, 1990) have emphasised the inherent flexibility of the so-called light-rail systems. They can, at least in theory, be developed in a phased way with each stage either seen as an end in itself or as a step to the next level of provision. Bonz (1990) argues that it is this which distinguishes the light rail from full-metro system although the definitions are by no means clear cut.

Conventional tram systems can be upgraded by increased track segregation, especially in congested areas, by phased progressive extension of the network and by the use of more modern, higher capacity vehicles (Catling, 1986). In practice, the scope for phased development may be limited once a particular technology has been put in place and the urban fabric has closed in on the system. Light rail has been able to provide considerable increases in capacity in some situations, Manila and Tunis have been cited as examples, without the disadvantages of providing full metros or segregated express bus systems. Light rail may also be seen as a way of extending or complementing existing metro systems in expanding urban areas.

The capital and operational costs of different rail systems vary greatly (Table 6.9) and will be influenced by local conditions (existing urban fabric and route network, terrain) and also the efficiency with which the system is managed. In general (Armstrong-Wright, 1986), buses provide the lowest system costs – from 0.02 to 0.08 US cents per passenger km compared with 0.03 to 0.10 for trams and 0.10 to 0.25 cents for different types of light rail and metro. However, for buses to provide comparable capacity requires exceptional road conditions and a level of bus segregation from other traffic that is rarely attainable. In a reserved bus lane in Bangkok a mixture of 250 standard and 150 mini-buses have accommodated 18,000 passengers per hour and in Bogota a mix of large and small buses on three lanes but with other traffic was able to deal with 32,000 passengers an hour but at slow speeds (Armstrong-Wright, 1986).

Possibly as a result of the perceived scale of their urban transport problem, many large cities are opting for expensive, rail-based transits. In Singapore – possibly not typical – the Mass Rail Transit is seen as a key component in a long-term plan to restructure the urban fabric and channel movement into a comprehensively planned, efficient land-use framework. It is, therefore, trying to anticipate urban transport problems whereas many cities are in a less enviable position and are having to insert a rail system into an existing congested urban fabric and transport system – 'neither easy nor optimal' (Figueroa and Henry, 1991).

Buenos Aires has a metro dating from 1913 but recently expanded and in 1969, Mexico City opened its metro. A number of other principal cities in the developing world have since adopted rail-based transits: São Paulo, Rio de Janeiro, Santiago, Caracas, Guadalajara, Salvador, Cairo, Helwan, Alexandria, Calcutta, Hong Kong, Singapore, Manila, Seoul, Taipei, Istanbul and Monterey. Moreover, there are similar plans for Belo Horizonte, Kuala Lumpur, Cordoba, Brasilia and Bogota, to mention just some cities. These rail-based transit systems cover the spectrum of technologies outlined above and have all been engineered to high standards and are well operated and efficient (Allport, 1991). Because of their high cost, rail transits, especially those of high capacity, can only have restricted coverage and most have been provided to serve corridors of high demand. Yet along these corridors buses are often still the dominant mode and the rail transit tends not to be well integrated into the wider pattern of demand (Figueroa and Henry, 1991). Only in Mexico City and Tunis has there been some attempt to remove the competition.

A number of common problems face developers of rail-transit systems (Allport, 1991). Problems of land acquisition, disputes over routes, changes in plans (sometimes related to frequent changes of government) and shortages of materials and funds have all contributed to the late completion of 75 per cent of all the projects, with an average overrun of two years and average completion times of seven years. This in itself accounts for inflationary cost increases and practically all the projects exceeded their budgets, sometimes dramatically. Calcutta's 10-mile metro opened late in 1995 at a cost of £300 million and some 12 years behind schedule.

Once in operation patronage has often been below forecast levels and the rail systems have failed to compete effectively with buses. The Hong Kong metro is one of the few which is likely to recover its costs and many have to be heavily subsidised with fare-box revenues falling far short of operating costs. Even so, fare levels are such that they are beyond the pocket of the poorest sections of the community and it is the middle- and higher-income groups that gain most benefit. The impact on traffic congestion has usually been minimal and the rail transits have done little to improve the overall quality of urban life or transport (Figueroa and Henry, 1991).

Between 1975 and 1978 two partially completed lines in Rio de Janeiro, with federal support, accounted for 57 per cent of all state public investment, but no private transport operators link into the rail routes and while operations on one line are 'satisfactory' on the other they are 'poor'. It can be argued that in this, as many other cases, the problems are institutional and managerial rather than technical but it is often these problems which are the most difficult to resolve in the context of Developing Countries.

Certain conditions need to be satisfied if rail transits are to be successful (Allport, 1991). They should be confined to arterial corridors on which peak bus flows approach 15,000 persons an hour before the metro is considered. Such corridors are unlikely to exist in cities of less than five million inhabitants or where movements are highly dispersed and not markedly linear in pattern. Successful metros are most likely in cities with per capita incomes in excess of $1,800 (or $1,500 at the national level) and massive in-migration tends to reduce average income levels and be detrimental to metro success. However, the metro allows the growth of activity in the city centre without restriction by declining accessibility but the overall impact is permissive rather than positive. It can certainly be argued that there are few cases in which the emphasis on high-cost, rail-based transport solutions have been wholly justified and there are many who would now urge caution on Developing Countries and the fullest consideration of all possible strategies.

## TRANSPORT STRATEGIES FOR URBAN AREAS

Wright (1992) has argued strongly that successful urban transport strategies must relate effectively the transport technology adopted, the characteristics of the urban area itself and the population that provides the demand. Dimitriou (1992) has made a similar case with specific reference to Indonesia. The emphasis has long been on technology and this has certainly been encouraged by external consultants, aid donors and equipment manufacturers. Paradoxically, it seems that it is easier to sell and fund high-technology products and there are powerful vested interests involved both within the Developing Countries and also outside to ensure that it stays this way. The World Bank has now concluded that too many of the projects put to them for support involve heavy capital outlay and continuing expenditure but provide little benefit to the mass of the people (Ridley, 1986) and a series of World bank technical papers have attempted to provide guidelines for examining the options (e.g. Armstrong-Wright, 1986).

More recently, based on wide experience in urban transport planning both in the United States and Latin America, Wright (1992) has pointed to the proved inadequacy of the existing planning models for dealing with urban transport problems and proposes an alternative in which transport projects and choices are viewed as the production and consumption of sets of 'characteristics'. Urban residents in general and transport users in particular derive utility from the characteristics of certain vehicles, modes or services. A given characteristic may be obtained from more than one mode, vehicle or service and each of these will exhibit numerous characteristics – a car has certain desirable characteristics (speed, comfort, convenience) and some undesirable (cost, environmental impact). Individual users and social groups will have different perceptions of the value of each characteristic (e.g. cost, frequency, convenience, time). In each situation there will be different evaluations of transport capacity, environmental impact, sustainability and public expense.

Emphasising as it does the unique features of each situation, the relevance of this approach to the problems of Developing Countries will be apparent because all too often there has been an unthinking transfer of technology which provides transport of a kind which does not relate to the characteristics of the city and its people. For example, the emphasis on the car and large-scale road engineering in Mexico City, São Paulo, Manila and Lagos, which have population densities two to three times that of cities like London or Atlanta, where space for vehicle circulation and parking is restricted and where the bulk of the population cannot hope to own a car, would hardly seem to be logical long-term transport planning or showing consideration for the vast majority of the population.

A transit option model which combines elements of the traditional World Bank guidelines and Wright approach is given in Figure 6.6. The important point to note in the context of Developing Countries is that in seeking more efficient urban transport strategies it is essential to give far greater weight than hitherto to

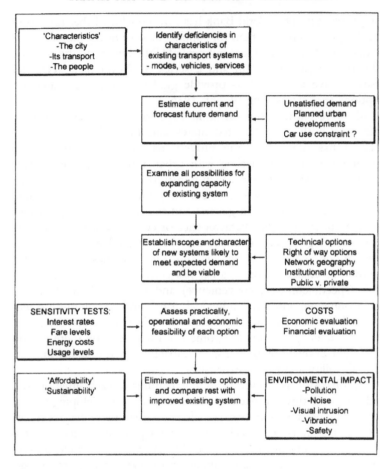

*Figure 6.6* Transit option study
*Source*: Based on Armstrong-Wright (1986), Wright (1992)

the urban and socio-economic characteristics of each city (Dimitriou, 1992). This is the vital starting point. It is then necessary to identify the deficiencies in the existing transport systems and whether these are technical or managerial in nature. Having forecast demand, the existing systems must be examined for every possible way of expanding capacity (e.g. changes in management or maintenance practice may ensure far greater bus availability) and only when all possibilities have been exhausted should attention be turned to new capacity options and each of these should then be

*Table 6.10* Transport systems and the environment

| | Air pollution | Noise | Visual intrusion | Safety |
|---|---|---|---|---|
| Walking | 5 | 5 | 5 | 3 |
| Cycling | 5 | 5 | 4 | 2 |
| Car | 1 | 2 | 2 | 2 |
| Bus, mixed traffic | 1–2 | 3 | 4 | 3 |
| Bus, reserved lane | 3 | 3 | 4 | 3 |
| Bus, bus ways | 4 | 4 | 4 | 4 |
| Tramway | 5 | 3 | 3 | 3 |
| Light rapid, surface | 5 | 3 | 3 | 4 |
| Rail rapid, surface | 5 | 2 | 2 | 4 |
| Rail rapid, elevated | 5 | 1–2 | 2 | 5 |
| Rail rapid, tunnel | 5 | 5 | 5 | 5 |

*Source*: After World Bank, (1986); Wright (1992)
*Notes*: 1 – very poor, 2 – poor, 3 – average, 4 – good, 5 – very good

scrutinised with respect to financial, technical and environmental factors (Abelson, 1995). Particularly, there is no point in developing options that are affordable neither by society nor by individual users and will thereby be rendered unsustainable with time – this can only result in a negative impact on the overall development process. It has been argued with respect to Chile, and it is true for many other Developing Countries, that scarcity of public finances is likely to be the principal determinant when considering transport efficiency improvements (Hall *et al.*, 1994) and this forces emphasis on proper evaluation and design.

It would be unwise in cities of Developing Countries to adopt any one option to the exclusion of all others and it would be premature to discount the continuing contribution of walking, HPUVs and IPT, although these could well be seen as complementary to higher technology options. In this way some cities might avoid becoming car-dominated in the future.

In cities world-wide, and perhaps especially in Developing Countries, little attention has hitherto been given to the broader, environmental impact of transport systems. There are now strong arguments for changing the emphasis and there is an opportunity in some Developing Countries to adopt and develop a combination of urban transport options (Table 6.10) which reduce to a minimum the less desirable effects of transport provision.

# 7

# SEAPORT 'GATEWAYS' AND PORT DEVELOPMENT

Gateway functions are concerned with exchange between the gateway settlement's home region or hinterland and external places which may be contiguous in a geographical sense or at considerable distance and possibly separated by maritime space (Bird, 1980, 1983). Located at the interface between land and maritime transport, the seaport is a clear and special case of the gateway, providing as it does the means by which a hinterland is connected to overseas trading areas – the foreland.

Many of the ports in what are now Developing Countries provide extreme examples of this gateway function, because they were the initial points of contact with a whole range of external influences – economic, political, social, and the bases from which these influences were diffused into the hinterland. It can, of course, be argued that many of these influences were malignant in nature – colonialism, dependency, underdevelopment – but what cannot be disputed is the profundity and the irreversible character of the changes initiated. The ports are often the largest centres of population, in many cases the capital city, usually the main concentrations of industry and therefore of interest as settlements. However, they become critical elements in the development process and significant influences on the modernisation of their hinterlands (Hoyle, 1988) and a case can be made that the stage of development in any territory is to a considerable degree a function of the capacity and sophistication of the available port facilities (Hoyle and Hilling, 1970; Hilling, 1990).

While the principal function of the port is the facilitation of exchange of goods between land and sea transport, as implied above, they are in fact multi-functional. As break-of-bulk points in the transport chain they are the logical location for industrial activity and with scale economies there has emerged the concept of the Maritime Industrial Development Area (MIDA) (Hoyle and Pinder, 1981; Hoyle and Hilling, 1984). Industrial development strengthens the links between the port and its region for which it becomes more than just a gateway. Both in themselves and in the economic activities they and their urban areas attract, ports have traditionally been great generators of employment. Because they are such significant growth poles, port planning should be a central plank in wider physical development planning strategies. Tema, Ghana, was created as part of wider regional development strategies (Hilling, 1966), San Pedro in Côte d'Ivoire was built to stimulate development in the country's previously isolated south-western region and in Malaysia, after 1970, Penang, Kuantan and Johore Baharu were seen as essential elements in regional growth policies (Fong, 1984). In 1993, Vietnam announced port developments at Cai Lan, Han La and Thi Vai to promote economic development in their regions and Thailand sees new port development at Laem Chabang, Songkhla and Phuket in a similar way.

## THE PORT 'SYSTEM'

Ports may conveniently be viewed as systems in the sense that they combine a number of distinct but separate elements which are linked into a functioning whole, which entity may operate without external influences (a 'closed' system) or, typically, in the case of ports, with external influences of great variety and complexity (Hoyle and Hilling, 1984). The way in which the elements are linked, the structure of the system, can be analysed, as can the way in which this changes over time and the manner in which the entity responds to external stimuli. Robinson (1976) has identified five levels of port system (Figure 7.1)

At the simplest level is the *intra-port system* which is all the activities taking place within the defined boundaries of the port – cargo handling, storage, clearance, ship servicing, etc. – and the organisations involved – ship owners, agents, port management, stevedores, customs, transport firms, etc. In reality, even this is a far from simple combination of elements. The *port-hinterland system* includes the transport services that bind the port with the land area it serves and the significance of the port rests in the link that it provides between this hinterland and shipping connections with overseas trading areas – the *port hinterland-foreland system*. Ports are

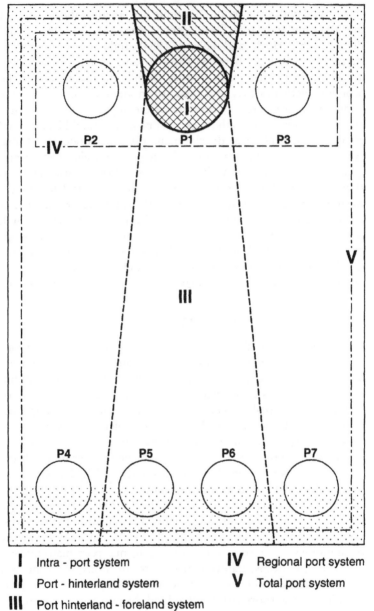

I  Intra - port system  IV  Regional port system

II  Port - hinterland system  V  Total port system

III  Port hinterland - foreland system

*Figure 7.1*  Port operating systems
*Source:*  Based on Robinson (1976)

often in competition for the trade of a common hinterland, the *regional port system*, and this may operate at different levels – ports on the same estuary (Tilbury and Sheerness on the Thames), ports in the same nation (Port Kelang and Johore Baharu in Malaysia) or internationally (Port Kelang, Singapore, Jakarta). The competitive relations of East Africa's seaports have been examined by Hoyle and Charlier (1995). At the most complex level each port is part of the *total port system* - the global maritime system.

It is implied above that ports are both complex systems and also systems of almost infinite openness, subject as they must be to influences of great variety at all the levels noted and often having little control over these variables. Arguably, the ports in Developing Countries have less control over their destinies than the longer-established ports of mature economies. The latter often have the advantage of long periods of evolution and slow adjustment to change and the build-up of managerial and labour skills, with access to levels of financial, technical and human resources that are lacking in the Developing Country. Not surprisingly, ports everywhere are dynamic entities and they respond to and reflect shifts in demand which derive from spatial and structural changes in both their hinterlands and forelands. With little or no control over the factors which influence these changes in demand for their services, it is not surprising that ports have great difficulty in achieving balance between that demand and the supply of facilities that they make available. At the level of day-to-day operations we shall return to this topic.

It will also be the case that the fortunes of particular ports will wax and wane in response to the exploitation of the hinterland. In the case of Nigeria, Ogundana (1972) identified a process of cyclical change with alternating periods of diffusion and concentration of port activity as individual ports responded to changing demand. Changing patterns of port activity over time are also a feature of the Taaffe *et al.*, (1963) transport and development model (Figure 7.2). In this the starting point is scattered port activity, each port's hinterland being restricted by the low capacity and range of traditional transport links (e.g. human porterage). Ports favoured by higher capacity penetration routes, perhaps along a river but more frequently by the construction of railways, will expand. The hinterlands of many ports have been determined by rail links – Lagos, Mombasa, Dar-es-Salaam provide early examples while more recently there have been Nouadhibou (Mauritania), Saldanha Bay

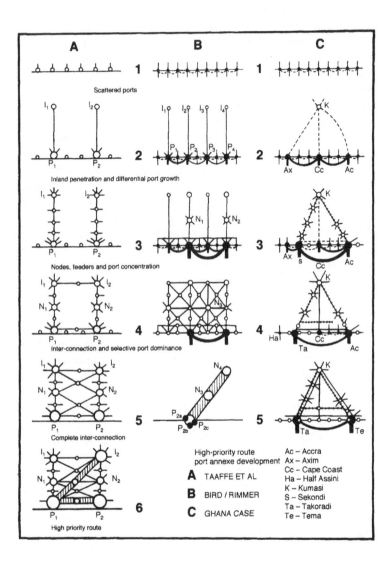

*Figure 7.2*   Models of port and transport development
*Source*: After Hilling (1977b)

(South Africa) Tuburao and Ponta da Madeira (Brazil) and Owendo (Gabon). In the later stages of the model some of the early ports disappear altogether as activity becomes concentrated at locations able to adapt to changing demand (Hilling, 1977b).

The Taaffe model ignores the crucial maritime connections but has been modified (Bird, 1971) to reflect different types of shipping itinerary (multi-port, selective, feeder) and the development of outports when existing ports reach capacity. It is also worth noting that the neat, historical continuity implied by the initial model does not always apply (Figure 7.2,c) as in Ghana, where Accra was able to progress from surf to lighterage port but was not suitable for the construction of deep-water berthage and a completely new location, Tema, had to be developed (Hilling, 1977b).

The ability of a port to respond to changing circumstances will in large measure reflect four aspects of its geography – its situation with respect to hinterland activities and to the sea lanes which serve international trade and its site conditions both on the land and sea side.

## SITE, SHIPPING TECHNOLOGY AND PORT MORPHOLOGY

While ports in particular regions may have some characteristics in common, by definition each port is unique and will comprise a distinctive combination of features relating to situation and site. The value of different coastal locations for the shelter that they provide for ocean-going ships and the ease with which landings can be effected or cargo transfer facilities provided will be determined by an amalgam of site factors (geomorphic, climatic, oceanographic), by the technology adopted for the ship–shore transfer and the engineering works that are required. One attempt to model the evolution of ports (Bird, 1963) is far from representing the 'Anyport' its name implies and is rather specifically a river port with a high tidal range – could certainly be London, less clearly Calcutta – but is not a low-tidal range, open-dock port like Rotterdam (Table 7.1). Even less can the model be applied to ports in very different geographical conditions such as the linear, surf-beaten coast of West Africa (Hilling, 1969a), the drowned estuaries, low-tidal range, coral fringed coast of East Africa (Hoyle, 1983) or the coral islands of the Pacific (Dunbar-Nobes, 1984). Even on a coast such as West Africa variations in local conditions (White, 1970)

*Table 7.1* The characteristics of 'Anyport'

| Original 'Anyport' | Australia | East Africa | West Africa |
|---|---|---|---|
| 1. Primitive era | 1. Lighterage | 1. Dhow traffic | 1. Surf port |
| 2. Marginal quay extension | 2. Marginal lighterage quay | 2. Lighterage quay | 2. Lighterage quay |
| 3. Marginal quay elaboration | 3. Finger piers | 3. Marginal quay extension | 3. Deep-water berthage |
| 4. Dock elaboration | 4. Marginal wharf provision | 4. Simple lineal quayage | |
| 5. Simple lineal quayage | | | |
| 6. Specialised quayage | | | |
| (Bird, 1972) | (Solomon, 1963) | (Hoyle, 1968) | (Hilling, 1977b) |

*Table 7.2*   West African coastal types and port development

| | Features | Stages | Examples |
|---|---|---|---|
| Guinea | Estuarine mangrove, sheltered anchorages, sheltered shore landing places | 1. Ship's boats<br>2. Lighterage piers<br>3. Deep-water marginal quays | Freetown<br>Banjul |
| Elmina | Small, rocky headlands, unsheltered anchorage, partially sheltered landing places | 1. Surf or ships' boats<br>2. Artificial lighterage harbour<br>3. Artificial deep-water harbour | Sekondi<br>Accra<br><br>Monrovia |
| Dahomey | Smooth, dune-backed coast, unsheltered anchorage, unsheltered beach landing places | 1. Surf port<br>2. Unsheltered lighterage pier<br>3. Artificial deep-water harbour | Nouakchott<br>Cotonou<br>Lomé<br>Kpémé |
| Lagos | Smooth, dune-backed, deep-water lagoons, natural ship access to lagoon (Lagos), artificial access to lagoon (Abidjan) | 1. Ships' boats<br>2. Lighterage piers, marginal quays<br>3. Deep-water marginal quays | Lagos<br>Abidjan |

have resulted in rather different types of port development (Table 7.2) and this will find echoes in parts of Latin America and Asia.

At many ports cargo handling initially is to and from sea-going vessels at anchor, whether in a river, sheltered roadstead or off an open coast, by small boats, whether belonging to the ship, of local style or in the form of barges. This is the process of 'lighterage' which is still the main, and sometimes the only method of handling cargo at some small river, coastal and island ports (Brookfield, 1984). Many of India's smaller ports are still of the 'roadstead' type with cargo handled by lighters – Navlakhi and Mandvi (Gulf of Kutch), Redi (Goa), Neendakara (Kerala) and Nagapattinam (Bay of Bengal) being just a few of them. Perhaps surprisingly, a port such as Hong Kong is still very dependent on lighterage with some 30 per cent of its great container throughput handled in this way.

In general this method of cargo handling is slow, of low productivity and may only be possible in daylight hours in some ports. There may be restrictions on the quantity and type of cargo that can be handled – large vehicles or project cargo may be impossible and bulk cargoes certainly difficult. Accra, Ghana, in 1961 handled

*Plate 7.1* The use of lighters in roadstead ports such as Malacca, Malaysia, involves double handling and restricts the volume and type of cargo that can be dealt with.

1.25 million tonnes by surf-boat from ships anchored off shore but this was an exceptional performance and limited to a unit size and weight that could be carried by at most a few men. Cotonou, Lomé and Nouakchott, before they were provided with deep-water berthage had annual capacities of no more than about 250,000 tonnes and given that Cotonou (Benin) and Lomé (Togo) were the only ports of their countries this was effectively the ceiling on seaborne trade through national ports and a severe restriction on trade and development potential.

If trade growth is not to be impeded such 'primitive' port facilities must be upgraded – in 1993 the Maldive Islands embarked on the construction of a deep-water quay to reduce lighterage operations. On unsheltered coasts the first development may well be a finger quay but rarely in such conditions are these suitable for ocean-going ships. Where the natural shelter is inadequate it has to be provided artificially by the construction of protective breakwaters – as at Madras (1890s), Visakhapatnam (1930), Takoradi (1928) and Tema (1960s). The length and character of such breakwaters will be determined by local conditions of climate (prevailing wind and strength, storm frequency) and sea (tidal streams, currents, waves – locally generated and long distance). Shelter for the sea-going ship becomes increasingly important with more sophisticated methods of cargo handling (e.g. placing of containers into the cell guides in holds, roll-on/roll-off) but breakwaters can be costly to construct and when completed provide an envelope for port development that may well prove restrictive with time (Hilling, 1977b). Great thought therefore has to be given to the design in relation to the services to be provided and potential growth of traffic (Wood, 1990).

Where there is a degree of natural shelter, as in rivers, estuaries or large bays, it may be possible to dispense with breakwaters and build either marginal quays, along the line of the shore, or finger quays, at right angles to the shore, for the sea-going ships. In the sheltered Lagos lagoon, the Freetown estuary and on Bangkok's Chao Phrya river, marginal quay provision is relatively simple and less costly, in general, than in situations where breakwaters have to be provided. It is also easier to gradually extend the marginal quays as demand requires.

In situations where there is a wide tidal range or where the river flow regime is variable it may be necessary to provide the deepwater berthage within enclosed docks where water levels can be maintained by impounding behind lock gates – Bombay and Calcutta provide obvious examples. This will be a costly solution and, as in the case of breakwaters, provides a finite envelope which, once filled, allows little scope for further expansion. The entry lock dimensions may also prove inadequate for larger ships and the enclosed dock may eventually have to supplemented by new facilities – as happened at Haldia for Calcutta and Nhava Sheva for Bombay. It is particularly important where enclosed docks have to be built that great attention is given at the design stage to lock dimensions, limiting water depths, ship turning circles and land availability to accommodate future changes in traffic.

The distinct advantage of the deep-water berth over lighterage is the elimination of double handling of cargo, increased productivity, faster handling and faster ship turnround. Lighterage may be adequate for small quantities of small items but as traffic increases or as bulk handling (e.g. grain, coal, ores, cement) becomes necessary so deep-water berthage becomes vital. When Ghana started to export manganese during the First World War it was through the lighterage port of Sekondi with manual loading and unloading of lighters on shore and at the ship and it was the inefficiency and cost of this which was a major factor influencing the decision to build, in the 1920s, the adjacent deep-water port at Takoradi. Even now, a cement factory at Takoradi is maintained by use of self-propelled barges for lightering of clinker from vessels at anchor, a decision never properly rationalised in view of the availability of deep-water berths and a conveyor system for bauxite that could be used. In India, iron ore is loaded from lighters at Redi, Goa, and clinker at Sikka, Gulf of Kutch.

A large number of ports have been built in response to the need to handle dry bulk cargoes – Saldanha Bay (South Africa), Buchanan (Liberia), Nouadhibou (Mauritania) and Tuburao (Brazil) are just a few of those for iron ore, Sepetiba (Brazil) and Tanjung Bara (Indonesia) for coal and Kpémé (Togo) for phosphates. The large size of the vessels in these trades makes special demands on port facilities, especially water depth, but in oil trades, while the vessel size may be even larger, the ship does not necessarily require a full quay structure and mooring dolphins may well be adequate with a bridge to take pipework and service vehicles to the ship's side. Indeed, it is possible to moor the ship by the bow to an off-shore Single Buoy Mooring (SBM) from which the connecting hoses can be picked up and this system is used both for loading (e.g. some Gulf terminals, off-shore Nigeria and Gabon) and discharging (e.g. Tetney, UK, and the Louisiana Off-shore Oil Pipeline).

Because vessel size has increased so greatly in bulk trades and also because large areas of land are required for storage (stock piles, tank farms) it is often necessary to locate bulk terminals away from the constricted traditional port areas and place them adjacent to the processing industries they serve – oil refineries, steel works, fertiliser factories and grain mills. In this way MIDAs emerge (Takel, 1974) with access to deep-water, plentiful land and good hinterland links. Richard's Bay (South Africa) was planned as a MIDA (Wiese, 1984) and Thailand is currently developing new industrial ports at Laem Chebang and Songhkla. Although not able to take large ships, Ghana's port of Tema was planned to serve specific industries based on imported, bulk raw materials – aluminium smelting, oil refining, steel making, cement making, grain milling and food processing.

The increasing use of containers and roll-on/roll-off for general cargo also places a premium on well protected, deep-water berthage with ample land available for parking. While Hong Kong demonstrates that large numbers of containers can be handled efficiently at moorings, this probably needs a sophistication of organisation and logistics which is lacking in most Developing Countries.

## CHANGING TECHNOLOGY OF MARITIME TRANSPORT

It has been suggested above that there is a close relationship between the evolving port, the characteristics of its site and shipping technology and it is worth considering the latter in more detail. Until the 1960s, specialised passenger liners and oil tankers apart, most dry cargo and indeed much passenger traffic was carried in general-purpose vessels of rather uniform design and size characteristics.

The typical cargo vessel, whether 'tramp' (plying by inducement) or 'liner' (regular, scheduled sailings between fixed ports) was 'tween decker (had several decks) from 6,000 to 15,000 deadweight tonnes (dwt), the effective carrying capacity, with lengths up to 150 m, beams of 14 to 20 m and draughts of 6 to 9 m. At most ports the access channels, lock dimensions and berths were related to ships with these parameters. Most of the vessels had their own cargo-handling gear and could be independent of shore-based equipment. On regular liner routes the vessels often had features appropriate to the trade on that route – deep tanks for vegetable oils and heavy-lift derricks for logs in West Africa, refrigerated capacity where there was fruit.

The cargo on shore and in the ship was handled by labour-intensive methods which were often slow and increasingly costly with time. In the 1950s shipowners sought more productive and cheaper methods and various forms of 'unitisation' were introduced. What these involved was the grouping of individual items into standardised units which allowed the maximum mechanisation of the handling process. This could mean 'packaging' (crating, strapping sawn timber), use of pallets, the use of containers, the movement of whole vehicles (roll-on/roll-off, RO/RO) and even the carriage of loaded barges (see Chapter 2 and Hilling, 1977a).

The effect of these changes is summarised in Table 7.3 and given the range of new specialised ship types that have come into use (Gardiner, 1992) there are obvious implications in terms of port provision. The conventional, fairly uniform, general cargo

*Table 7.3*   Cargo-handling technology

| Conventional | Improved |
|---|---|
| Labour intensive | Capital intensive |
| Low level of mechanisation | Highly mechanised |
| Low man/gang/berth productivity | High man/gang/berth productivity |
| Small ships | Larger ships |
| Many ships | Fewer ships |
| Uniform ship type | Specialised ship types |
| Numerous uniform berths | Fewer specialised berths |
| Port dispersion | Port concentration |

berth may have to be replaced by container berths, RO/RO berths, specialised pallet handling facilities and possibly mechanised hand-ling of traditional cargo (e.g. bagged cocoa, coffee). In the bulk trades economies of scale have led to increased ship size with tankers of over 500,000 dwt and bulk carriers of up to 360,000 dwt. Few ports can accommodate these giants and they must be specially equipped.

If the technological gap between the advanced and Developing Countries is not to widen, critical investment decisions have to be made. In the iron ore trade, Liberia and Mauritania are at a disadvantage in having 1960s-vintage ports with limits of about 120,000 dwt whereas Ponta da Madeira, Brazil, is able to take a 350,000 dwt vessel. In the coal trades vessel sizes are rather lower (Browne and Hilling, 1992) but in Indonesia, Tanjung Bara can take 180,000 dwt vessels and Puerto Bolivar, Colombia, 150,000 tonners.

It is often the case that such specialised mineral handling terminals are financed as a part of the mining company's develop-ment package, including, where necessary, the hinterland transport links. This is not usually the case with general cargo facilities, which have to be financed by the port or the government. It is for this reason that in many Developing Countries ports have been slow to adopt the new technology but there may be reasons other than purely financial that come into play. A throughput of 500,000 tonnes a year has been suggested as a minimum to justify the cost of a specialised container berth and many Developing Coun-tries do not have this tonnage of general cargo and much trade is of logs, bulks, bagged or baled commodities, which are not suitable for container movement. The containerisable cargo may be ser-

*Plate 7.2*   Cuban bulk sugar, hand bagged for unloading at Alexandria – a labour intensive, slow method of cargo handling which characterises many Third World ports (Photo: Cargill plc).

iously imbalanced (Hilling, 1973) – a high proportion of the containers to Nigeria return empty northbound. Containerisation makes heavy demands on scarce capital and reduces by as much as 90 per cent the demand for labour in situations where employment generation should be emphasised (Hilling, 1983). It is often the case that the inland transport is inadequate for the onward movement of containers.

Containerisation may be a factor in the progressive marginalisation of regions within countries or even of the countries themselves. As vessel size increases operational economics suggest fewer port calls – some round-the-world services call at as little as ten ports. There is the tendency for pivotal, main ports ('load centres' or hubs) to emerge, from which secondary ports are served by transhipment and feeder services (Hayuth, 1981; Marti, 1988). It is also the case that not all ports can accommodate the larger ships and for a variety of reasons, physical and financial, may not be able to adapt to the new demands.

Concentration of activity has been an inevitable outcome of containerisation and occurs at several levels. At many ports only

251

parts of the existing facilities may be suitable for the new technology – at Lagos container handling is concentrated on specially built berths at Apapa. At a different level, whereas Lagos handles 65 per cent of Nigeria's general cargo it handles 91 per cent of the containers (Hilling, 1983).

At a regional scale, Singapore has emerged as a major hub whereas Calcutta handles few containers directly and receives most by way of transhipment at Singapore or Colombo. Whether or not a particular port will receive its containers directly will be determined by the traffic generated and the amount of deviation from a main route for the ship to make a call – the larger the ship the more containers there would have to be (Gilman, 1991).

Kingston, Jamaica, has a location which suggests possible hub status in the Caribbean and has been used as such by several shipping companies. In 1993, Point Lisas Port, Trinidad, completed a container berth in the hope that a transhipment role might develop for trade between South American ports and markets in Central America and the southern United States. Colombo, Sri Lanka, and Dubai, United Arab Emirates, are also emerging as significant hub ports.

A hierarchical port system is emerging in response to the new technology and the Developing Countries tend not to be well represented at the top of the rankings. South Africa apart, Africa's top container port is 76th in the world ranking while South America's largest, Santos, Brazil, is ranked 36th (*Containerisation International*, 1991). Asia's Newly Industrialised Countries have ports well up the ranking; Hong Kong – 1, Singapore – 2, Kaohsiung – 4, Busan – 6, Bangkok – 21 (Comtois, 1994). Overall, Developing Countries account for only 12–15 per cent of world container trade and most are still making only tentative first steps in that direction, with many of the containers they do handle being by way of combo ships (conventional with some containers on deck) or small feeder vessels rather than large container ships. As industrialisation proceeds, so will containerisable cargo increase.

Massive congestion in many ports in Developing Countries in the 1970s pointed to the quick turn-round provided by RO/RO ships as a useful way of increasing port capacity. Increasing amounts of project cargo (machinery, equipment) and vehicle imports pointed in the same direction. In response to congestion at Lagos in 1974–7 over 20 per cent of shippers from Europe switched to RO/RO (Dickinson, 1984) and stern ramp RO/RO berths were incorporated into the new Tin Can Island port scheme. In practice, most newer RO/RO ships have slewing, quarter ramps and, although in general these tend to be rather larger vessels than those with stern ramps, they do not require special berths.

It may be that other technologies are appropriate where cargo volumes are small and handling places dispersed and without facilities and Brookfield (1984) has described the possible use in island regions of smaller barge carriers, landing craft that can run up on to beaches and amphibious lighters which could even carry small containers.

Undoubtedly in this new technology one can see the possibly undesirable pressure of external forces – shipowners and large multinational trading companies concerned more for their own good than the general benefit of the Developing Country. Each Developing Country has to decide whether or not to adopt the new technology, the balance between advantages and disadvantages (Hughes, 1977) and the possible consequences of being further marginalised.

## PORT CHARACTERISTICS OF DEVELOPING COUNTRIES

In a real sense every port is unique and ports can be found on all levels on the technological continuum yet there is also a general consensus that while they may vary greatly, the ports of Developing Countries nevertheless share some common characteristics (Nagorski, 1971).

For many, a colonial past is a factor of critical importance. This will be most potent for those many countries of Africa and Asia which obtained their political independence in the last 40 years but is certainly not absent in areas of much earlier independence in South America. In territories under colonial control, infrastructure, including ports, was provided only under extreme pressure and supply was invariably way behind demand. As the demands increased with independence – growing export bases, increased consumer and capital goods imports, industrialisation, infrastructure development – so the inherited port system was found seriously inadequate.

There was an urgent need for provision of additional port facilities but this was against the background of the rapid technological change already described, the danger of mistakes in technology selection (stern loading RO/RO berths at Lagos's Tin Can Island port development were obsolete by the time they were ready), the need to import many of the personnel required to design and construct facilities and most or all of the material and equipment necessary and the fact that it usually had to be done with borrowed money. There was also considerable external pressure both from the shipowners and also those involved in the provision of new facilities (consultants, construction companies) to go for technology and designs not always closely in accord with local conditions and real as distinct from perceived need.

Many of the Developing Countries have trade structures which have also been termed 'colonial' with the exports restricted to a small range of primary products – agricultural, mineral, forestry – and with imports covering the whole range of consumer and producer goods. This results in a directional imbalance in the quantity and quality of trade and often also in the seasonal demand for shipping services and port facilities (Hilling, 1973). Given the very different character of the commodities involved it may also be the case that quite separate handling facilities are needed – a conveyor system for loading bagged cocoa at Tema, Ghana, can effectively be used for no other cargo. Bulk handling plant is invariably dedicated to a specific cargo and has little flexibility in use. The Developing Country may well have to provide a considerable variety of cargo handling facilities in relation to the total trade.

The trade of most advanced, industrialised economies tends to be spread fairly evenly over the year but for many tropical Developing Countries there is marked seasonality (Hilling, 1973). In the provision of any transport services temporal fluctuations in demand, whether daily to cope with rush hours or annual as for ports, make it difficult to assess optimal supply of services and provide them economically. For countries dependent on agricultural exports (soya, cocoa, coffee, groundnuts) such seasonality is likely to be very marked – The Gambia's groundnut season is compressed into about four months. However, heavy seasonal rains often cause operating problems on roads, railways and waterways and may disrupt the flow of other commodities to ports – timber and even ores in some cases. Heavy rain (e.g. Asian and

West African monsoon seasons) can also cause severe disruption to cargo handling at ports – this may have to stop for long periods of time, hatches have to be covered and working hours are lost (Figure 7.3). This adds to the cost of the ship in port.

Faced with severe foreign-exchange shortages, Developing Countries in some cases impose import quotas and the timing of quota allocation can result in bunching of imports e.g. at the start of

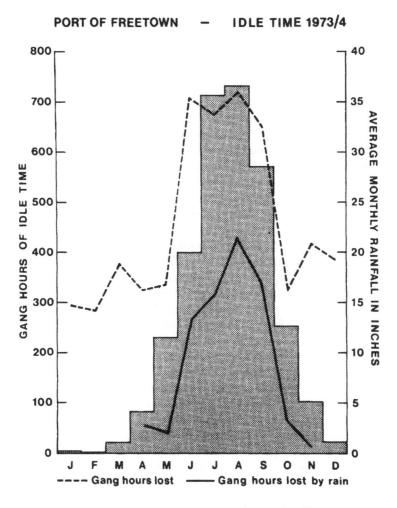

*Figure 7.3* Freetown - weather and cargo handling

a new round or sometimes at the end of a round as quotas have to be used up. As in many other countries, there may be surges in demand related to particular festivities in the calendar – Christmas is an obvious one – but even Mardi Gras or extended carnivals could have the same effect if only because all work has to be cleared in anticipation that it may not resume for some time!

While these variations in demand may be considered normal there are also the exceptional surges brought about by relief food aid as in many African ports in the mid-1970s and again in the mid-1980s. Rather than facilities in general it is often particular items of equipment which are not available. The high ambient temperatures and relative humidities together with torrential rain adversely affects equipment, especially where it has not been suitably 'tropicalised' – there have been cases of electric motors of wholly unsatisfactory design and crane cabs that could not keep out tropical rain. Shortage of money, lack of spares and poor maintenance in many Developing Countries ensure that all too much equipment is out of service at any time. In Nigeria there has been a continuing dispute over port charges, which include the use of equipment which all too often is not available and has to be provided by shipping companies at additional cost. For a time in 1991, Karachi's two pilot boats were out of action and no ships could enter or leave port.

Poor maintenance is all too common and the 'crumbling infra-structure' at Argentine ports (*Lloyd's List*, 10/11/1990), 'ageing' Chilean ports (*Fairplay*, 13/5/1993) and 'antiquated port equipment and bureaucracy' at Brazilian ports (*Lloyd's List*, 10/4/1991) are all descriptions that could be equally applied to many other Latin American and African and some Asian ports. Most of the ports are in the public sector, decision making is politicised and manage-ment short-comings reflect the inability to attract the best staff. The work force comprises far too many unskilled workers in ageing, closed groups and skilled operatives are in short supply. Bombay could apparently manage on half its present labour force of 40,000, Montevideo could reduce its 5,000 to 1,000 and at all South American ports the labour system is thought to be at the root of the port problem (*Lloyd's List*, 10/4/1991). The labour unions both in South America and India are powerful and inflex-ible.

The most used words in port literature are 'modernisation' and 'rehabilitation' , with some of the latter necessary at ports which are by no means old by any standards. For example, the World Bank has financed the upgrading of the electrical systems and warehousing at Tema, Ghana, and also, at both Tema and Takoradi, there has been a programme of wreck removal. In the Philippines there has been rehabilitation at Manila, Cebu, Iloila, Cagayan de Oro and Zanboanga, in Cochin the cargo systems have been modernised, in Bombay old quays reconstructed and in Ilo, Peru, there is an upgrading programme.

## THE PORT CONGESTION PROBLEM

For a variety of reasons the temporal flow of cargo in many Developing Countries is likely to be uneven with alternating periods when inadequate facililities are unable to cope with demand and periods of under-utilisation and idle capacity (Figure 7.4). It follows that for many of these countries port congestion, either seasonal or sometimes permanent, is a characteristic feature whereas for the ports of mature economies this is relatively rare and exceptional rather than normal.

This is certainly not a new phenomenon – one has visions from the last century of windjammers waiting in large numbers at Chilean nitrate ports and in the 1950s, the United Africa Company was so concerned with delays to shipping in West African ports that it conducted a survey to identify the causes (United Africa Company, 1957; Hilling, 1976b). The principal factors were rapidly increasing trade, increases in ship size in relation to port facilities, the inefficiencies of the land-side transport and climate.

Following the mid-1970s oil price increases the producers went on a spending spree and there was massive port congestion in Gulf and Nigerian ports. In the case of Nigeria, the congestion of 1975–6 has become part of maritime folk lore, with delays of 120 to 180 days not uncommon, crews either taken off ships at anchor or in desperate conditions and at a peak over 400 ships waiting to enter port. During the first half of the 1970s in a selection of 30 ports in Developing Countries the average waiting days per ship increased from 2.2 (1971) to 40.5 (1976) and UNCTAD set up an expert commission to assess the problem (UNCTAD, 1976). In 1985 Nigeria again had problems, in 1992 'unprecedented delays' and mounting costs were reported at Santos, Brazil, and in 1993 Ghana had a log-jam of uncleared goods, Klong Toey (Bangkok) had to reduce the number of ships permitted to call and in Nigeria a

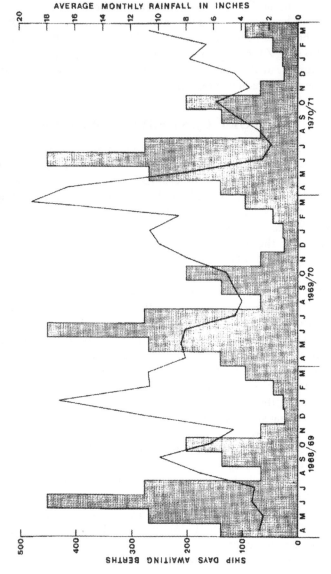

*Figure 7.4* Lagos – seasonality and delay

Presidential Task Force was set up to decongest ports. At many Chinese ports congestion has become endemic and in 1995 Bangladeshi ports faced severe congestion.

It is with some justification that it can be claimed that congestion on a large scale is a normal feature and the most obvious shortcoming of ports in Developing Countries (Arlt, 1989). Interestingly, in advancing reasons for this, Arlt includes all the factors identified in the 1950s by the United Africa Company – a graphic illustration that in such countries, *'plus ça change, plus c'est la même chose'*. It was also being argued that this port congestion was adding greatly to the cost of imports and reducing the competitiveness of exports and therefore seriously impeding and increasing the overall cost of general economic development (UNCTAD, 1976). When ships are held up for long periods at ports there is a reduction in ship earning capacity and an increase in costs and the shipowners usually impose a surcharge to compensate. This is passed on to the shipper but this was not done consistently so that some countries suffered more than others and UNCTAD (1980) proposed a more uniform system of charging.

Rapid growth in trade following the opening up of China's economy placed the ports under increasing pressure and in recent years congestion has become the norm. Many of the existing berths were for smaller ships and not suitable for the larger vessels which now dominate trade in both general and bulk trades. In 1993 China had 29 major port development projects in progress including new ports at Simao on the Lancong River in southwest China and Dayaowan in Dalien where four berths were completed in that year with six more to follow. At most other ports the number of berths was being increased to reduce congestion and often with upgrading to a 10,000 dwt minimum vessel standard, a modest, if not inadequate, standard.

In Nigeria, the response to congestion was a massive programme of port development. In 1979, a new 10-berth port was opened at Tin Can Island to serve Lagos, a turn-key project completed in only 15 months, and new ports were built at Calabar, Sapele and Warri. There has been only slow progress on a 50-berth new port at Onne, near Port Harcourt, and recession in recent years, while it has not entirely eliminated occasional congestion, has probably reduced the need for so many new berths (see discussion below). In Mexico, new ports are being built on the Pacific coast at Topolobampo and Pichilinque. In India, Bombay has a long history of congestion and the port of Nhava Sheva has been built to supplement its capacity and a satellite port for Madras is planned at Enncore. In Pakistan there are plans for a new port at Gwadar and in a purposeful attempt to divert traffic away from congested Bangkok, Thailand has invested in new port capacity at Laem Chabang, Mai Ta Phut, Bang Saphaan and Songkhla. In response to increasing demand from traditional dhow traffic a new port is being built at Dubai.

At the same time that congested existing facilities have needed expansion and upgrading there has also been the demand to adopt the new technology and an almost frantic attempt to provide the necessary facilities, especially for the handling of containers. In small numbers these can be handled with conventional ships and lifting gear or by cellular container ships with their own cranes but as numbers increase a shore gantry crane becomes essential and berths have to be provided with additional storage space – in 1992, Venezuela still had no container gantry crane and Ghana still has none.

By the 1990 Mexico's main ports had been provided with dedicated container berths and in 1980 China opened its first container berth at Tianjian, where there are now seven berths. By 1996 Xiamen will have two container berths and at Shekou, south China, a new container port is being built and there are plans for container facilities at Jintang, Ningpo and Daxie Island in the Shanghai area, Yantian and Ping Hai Wan. Much of this development is aimed at reducing dependence on Hong Kong, which now handles much of China's container trade by transhipment.

The Indonesian government is going through an aggressive phase of container berth construction at Tanjong Priok (Jakarta) and Surabaya with plans for a container terminal on Batam Island which, it is hoped, will take present transhipment trade away from Singapore and develop as a regional hub. In the Philippines it is thought that 75 to 80 per cent of the main port traffic will be containerised over the next 10 to 12 years and this will require considerable expansion of handling facilities. The Pacific Rim countries in general are showing a higher rate of container growth than any other world region (Turner, 1991) and any impediment to that growth will inhibit economic growth in general.

*Plate 7.3* The use of 'self-sustaining' container ships makes costly port container cranes unnecessary but is only suitable when the number of containers is small.

## PORT EFFICIENCY, PRODUCTIVITY AND CAPACITY

The intimate connection between port efficiency, productivity and capacity has all too frequently been ignored and it is reasonable to address this problem in more detail. The idea that there is rarely a balance between the demand for and supply of port facilities has already been noted and it is particularly marked in port operations in Developing Countries. Any measure of port capacity will be theoretical and notional but will depend on the facilities available and, perhaps of greater significance, the efficiency with which they are operated. For any item of equipment (pump, conveyor, fork-lift truck, crane) there will be a rated performance based on uninterrupted operation under specified, usually optimal, conditions. Invariably, the effective or actual performance is lower, and in the case of Developing Countries often very much lower, simply because the operating conditions are far below optimal – poor

261

management, poor maintenance, weather conditions and unskilled labour.

Within the total port system the berth is the critical sub-system – this is where the modal interchange takes place, the port's *raison d'être* (Figure 7.5). The capacity of a port will be the aggregate of the capacities of the individual berths and will be dependent on two aspects of productivity – the rate at which the cargo is handled between ship and shore (the ship cargo-handling system) and the rate at which the cargo arrives at or can be delivered away from the immediate berth area (the shore handling system). In the case of

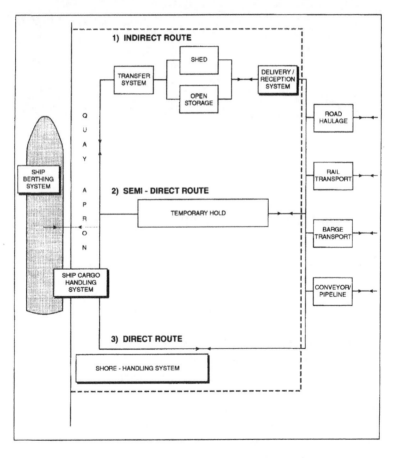

*Figure 7.5* The port-berth system
*Source*: Modified after UNCTAD (1973)

import cargo as long as the rate of off-loading is less than the rate of delivery the unloading can proceed but if the delivery rate is lower then cargo will accumulate in the berth area, unloading will slow down and when storage capacity is full may even stop. This is the classic congestion situation and leads to declining service and mounting costs to shipowners (Janssen and Shneerson, 1982).

Usually, cargo is off-loaded into storage (the 'indirect' route) and the amount of available storage becomes the critical factor. Often because it was designed for smaller vessels or slower systems of cargo handling, the storage in many ports of Developing Countries is insufficient and frequently is used wrongly, by consignees, as long-term storage rather than for the short-term transit warehousing function for which it was intended – government departments and agencies often being the worst offenders. In August 1993, cargo was reported as 'heaped mountain high along the docks while incoming ships were stuck for days awaiting their turn to unload' at a number of Chinese ports (*Lloyd's List*, 20/8/1993), while a little earlier Ghana's importers and their clearing agents were blamed for delay at Tema (*West Africa*, 21–7 June 1993). Guidelines to the amount of storage needed under particular conditions have been provided by Imakita (1978) and UNCTAD (1973, 1985).

Ideally, cargo is handled directly from the ship to onward transport or at most held only briefly – the direct and semi-direct routes (Figure 7.5). This is easier where the customs clearance is being effected at an inland location – a 'dry port' or inland customs depot such as Nairobi (Kenya), Kaduna (Nigeria) or Delhi (India) – or where the homogeneity of the cargo makes it easier – such as bulk cargo for single consignee.

In each situation the availability and efficiency of inland transport is critical and in many Developing Countries a restrictive factor – the necessary transport is not available when required. Of Djibouti it has been said that 'there is no infrastructure beyond the port' (*Fairplay*, 12/8/1993) and the hope that endemic congestion at Bangkok's Klong Toey port would be relieved by the opening of new facilities at Laem Chabang has been delayed by the inadequacy of the road and rail links to the new port. Disruptions to inland transport are an all too frequent feature in areas with unsurfaced roads and railways built to low design standards (Chapters 3 and 5) and rainfall heavy and seasonally concentrated . All too few routes are capable of maintaining the continuous, high-density movement required to sustain high levels of port activity. This problem can be illustrated by a specific study (Hilling, 1985).

The basic port function is everywhere the same – the efficient, speedy, safe and low-cost movement of freight across the land–sea interface. A key measure of the port's success will be the turnround time for the ship at the berth and this can be monitored, although rarely has been, by an analysis of the Statement of Fact or Working Programme produced by the ship's agent or stevedoring company. This shows how time is spent while the ship is in port and can be divided (Figure 7.6) into waiting time (delays before berthing) and time on berth. The latter can be divided into operational and non-operational time (when work is not scheduled) and the operational time includes actual cargo working and also time when, for a variety of reasons, no cargo handling is taking place – idle time. When recorded in sufficient detail, and it is important that they should be, the reasons for non-productive time can be identified and quantified and the bottlenecks become apparent.

An analysis of port time based on a Working Programme for a particular ship at a West African port (Table 7.4) reveals that rain, waiting for transport and administrative delays all contributed to a very low productivity of only 42 tonnes per working hour or 17 tonnes per hour that the ship was on the berth. A main problem

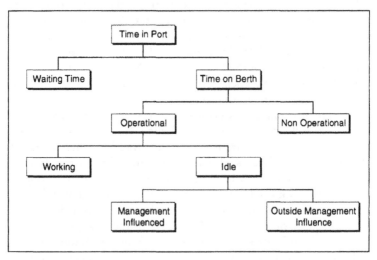

*Figure 7.6*  Ship time in port

*Table 7.4*   Analysis of ship working programme

| Ship: NOTIS | Arrived at port: Hours | Sailed from port: | |
|---|---|---|---|
| | | Percentage of time in port | Percentage of time on berth |
| Total time in port | 865 | | |
| waiting | 133.5 | 15.4 | |
| on berth | 731.5 | 84.6 | |
| Time on berth | | | |
| Non-operational | 174.5 | 20.2 | 23.8 |
| Operational | 557.0 | 64.4 | 76.2 |
| (a) working | 295.0 | 34.2 | 40.3 |
| (b) idle | 261.5 | 30.2 | 35.7 |
| rain | 55.5 | 6.4 | 7.6 |
| awaiting lorries | 91.0 | 11.1 | 13.1 |
| awaiting barges | 43.5 | 5.0 | 5.9 |
| administrative | 66.5 | 7.7 | 9.1 |
| Tonnage handled (carton milk): | 12,500 | | |
| Tonnes per working hour: | 42.37 | | |
| Tonnes per hour on berth: | 17.1 | | |

*Source*: Hilling (1985)

was the lack of onward transport, a factor over which the port itself has no control and this points to the need to see port planning in the widest context – there is no point in providing additional port capacity if the inland transport links of the port are neglected. The situation illustrated in this example is by no means unrepresentative of that prevailing in many Developing Countries and it is clearly important to improve the productivity and thereby the overall capacity of the ports.

## PORT PLANNING

Until recently it would have been true for most ports that they expanded in response to demand, often only when pressure had built up to an unacceptable level and in a piecemeal and unplanned way. There is a basic problem in that the optimal capacity for the shipowner is one in which his ship can immediately get alongside and start cargo handling but this implies spare capacity and a lower level of berth occupancy than is ideal from the port's point of view. This can be expressed in cost terms (Figure 7.7) – the lowest, total cost per tonne of cargo (B) being achieved at a lower level of throughput than that giving the lowest cost to the port (A). The lowest cost is a function of the relationship between berth occupancy and ship waiting time.

In response to large-scale congestion in the late 1970s there was unprecedented port development in the Arabian Gulf and Nigeria but much of the capacity provided at that time has proved to be superfluous, obsolete or wrongly located – the RO/RO berths at Tin Can Island have been mentioned above. Strategies adopted to deal with crises are unlikely to provide sustainable long-term solutions and more systematic planning is called for – but this requires a statistical base which is often lacking, patchy or inaccurate and an ability to forecast which is everywhere difficult but possibly excessively so for complex port systems in Developing Countries. Developing Countries cannot afford mistakes and there are strong arguments in favour of a national ports plan (UNCTAD, 1985) so that unnecessary competition can be eliminated, over-investment reduced and use of all facilities optimised – a possible strategy is illustrated in Table 7.5.

All too often port expansion is planned when the existing facilities are working below theoretical and acceptable levels,

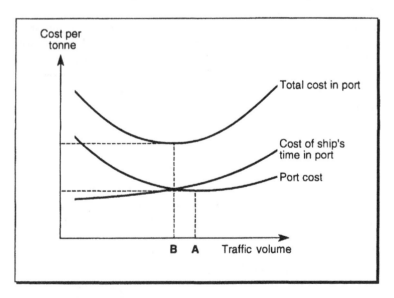

*Figure 7.7*   Port cost and increasing traffic
*Source*: From UNCTAD (1985)

*Table 7.5*   Port development strategies

SHORT-TERM (independent of longer-term planning)
Actions to improve management, identification of bottlenecks in existing system, maximise productivity of existing facilities/operations (*first priority*), identification of potential for expansion of capacity, establish satisfactory data base.

LONGER-TERM
*Establish development framework or national ports plan* – industrial, agricultural and mining sector plans, national pattern of consumption, international transit traffic, coastal conditions (geological, hydrographic), regional development policies/planning
*Develop Master Plan for each port* – define role of each port (trade? hinterland? transhipment? industry? transit? regional role?), define responsibilities of port (seaward side – conservancy, engineering, dredging; landward side – storage, transport), land use policy for port (freedom to acquire, control and determine land use), financial responsibility of port (commercial/profit making, self-financing, level of government support, decision-making responsibility, need for government approval, freedom to set rates and charges, need for flexibility, emphasis on sustainability

MEDIUM TERM (identifiable separate projects)
*Identification of projects* – consistency with long term plans (national and port), justification, feasibility, environmental assessment of options, operational analysis of alternatives, cost/benefit analysis, funding options

Modified from UNCTAD (1985)

what has been called the Russian approach – 'the ports don't work, so let's build some more' (*Fairplay*, 21/10/1993). The obvious, but often ignored starting point should be to maximise the productivity of the existing berths and only then to consider costly additional capacity. Demand from all sectors of the economy must be evaluated and there will have to be consideration of how the goods will be packaged, handled and carried – the technology of maritime transport. To translate this into ship calls requires estimates of productivity but from the example above the actual productivity at any time may not be the best guide. For example, for a forecast traffic of one million tonnes a year, 20 berths would be needed at a productivity of 50,000 tonnes per berth per year (a low productivity) but only 10 at 100,000 tonnes (reasonably acceptable level) – and at a good throughput of 150,000 tonnes only six or seven berths. This shows how critical the actual productivity can be.

In order to simplify the port planning in Developing Countries, UNCTAD (1985) has produced a series of planning charts which allow the port to calculate the required berths, storage and handling capacity for different types of cargo (general, container, RO/RO, bulk) under varying assumptions regarding working practice and productivity. One example of the many will be seen in Figure 7.8. The fraught question of productivity is the starting point, here expressed as average tonnes per gang hour, and it then moves through working practices to a figure of number of berths needed.

Existing ports should have the statistical base to plan extensions but facilities for a new type of traffic (e.g. container) or plans for a completely new port may require the planner to adopt productivity figures from another port. In this situation a comparable port should be taken – it is not reasonable to take productivity figures from, say Rotterdam, for a new port in a completely different

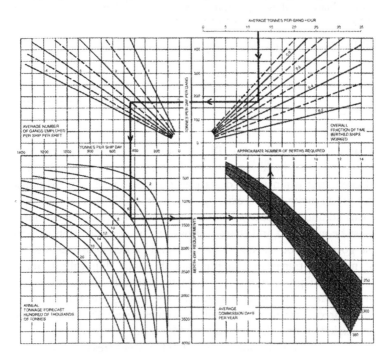

*Figure 7.8* UNCTAD berth planning chart
*Source:* From UNCTAD (1985)

environment (in every sense – climatic, terrain, political, social, economic) in a Developing Country. This is bound to distort the planning model and lead to over- or under-provision. It is certainly desirable to build some flexibility into designs and capacity.

There are dangers and cost implications either in delay in providing new facilities or in providing too much too soon and it has been suggested that the *whether*, *when* and *where* of port investment should be based on the technique of dynamic programming (Shneerson, 1981) in which economic criteria are used to optimise provision over time. In practice, it is probably the case that most port planning in Developing Countries is still a mixture of crisis management and a short-term approach with too little regard for the integration of the functional components of the system – the ship, the berth, the landward links (Figure 7.9).

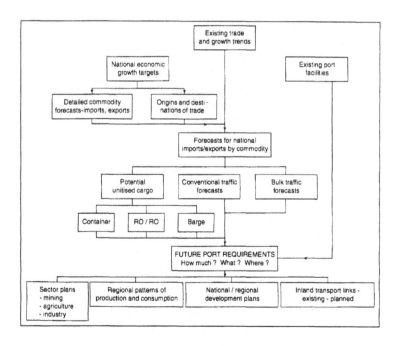

*Figure 7.9* A port planning model
*Source*: From UNCTAD (1985)

## PORTS – PUBLIC OR PRIVATE?

In many Developing Countries there is a strong tradition of public, central government involvement in the provision and control of ports and in terms of their efficiency many of the ports leave a great deal to be desired. Nevertheless it is too simple to argue that this is a causal relationship – two of the most efficient ports in the world, Singapore and Rotterdam, are in the public domain – Singapore through central and Rotterdam through municipal government. However, there has been growing intensity in the public versus private debate with regard to port control, as in many other areas of government involvement.

In many colonial territories the ports and railways were developed by government and continued to be controlled by government after independence. This was generally thought necessary for infrastructure, which it was considered would not attract private investment, but was reinforced by the often left-of-centre political complexion of many of the new governments. Yet under government control ports are in competition for scarce resources with what many might consider more deserving sectors of social and economic investment (Chiu and Chu, 1984). In this situation the ports have not always been able to command the finances they needed for capital investment and routine maintenance and this must be a significant factor in their inadequacy to meet demand. In 1991 the Argentine Administracione General de Puertos was taken out of the Ministry of Public Works and put into the Economic Ministry 'along with a multitude of concerns each striving for recognition . . . hard to convince that ports are important' (*Fairplay*, 23/5/1991).

In practice, port authorities span the spectrum from those which are wholly private (Figure 7.10) to those in which there is public involvement, the nature and directness of this varying greatly between countries and often changing over time (Taylor, 1984). In China, the Ministry of Communications is responsible for ports, although since 1984 there has been some decentralisation of control to municipalities, often with a vice-mayor as director. In India the main ports are controlled by trusts responsible to the Ministry of Transport, the chairman of the trust usually being a civil servant, but any changes (e.g. recent plans for an incentive scheme to boost productivity at Nhava Sheva) have to go through the time-consuming process of government approval. In South

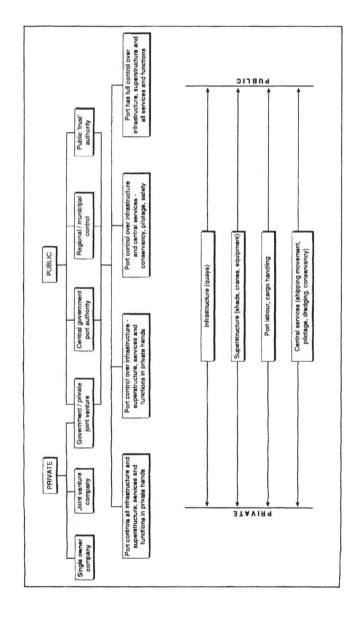

*Figure 7.10* Forms of port control

America government control has invariably been direct and some-times stultifying – AGP in Argentina, PORTOBRAS in Brazil, EMPORCHI in Chile, Puertos Mexicanos in Mexico and ENAPU in Peru. In virtually all African countries, South Africa included, ports are government controlled.

The call for privatisation has been prompted either by economic success, as in the Pacific Rim countries including China (Arrow-smith, 1995), or by the perceived inadequacies of public control in most African and South American countries. The pace and form of privatisation varies between countries but least progress has been made in Africa, although in 1992 the Nigerian Port Authority was transformed into a commercial entity, Nigeria Ports plc, with power to capitalise on assets, borrow money and fix charges and with responsibility for recruitment, training and material procure-ment devolved to three port 'zones'. In Egypt the El Dekleila Container Terminal is operated by a private consortium with government participation.

*Plate 7.4*   At Yantian, China is developing one of a number of new ports with high-speed road and rail links for conventional cargo (right) and containers (centre).

Since 1991 there has been limited privatisation in India mainly by contracting out selected, specialised functions (e.g. shipping line provision of equipment on dedicated berths, container freight stations) – a move towards Hong Kong's 'free-fiscal' rather than Singapore's 'economic-management' approach. In Calcutta in 1992 a berth was leased to Tata Iron and Steel and at Nhava Sheva the Maersk Line is providing a container freight station. At Madras, port labour has opposed proposals for the privatisation of specialised handling, operation of floating cranes, pilotage and dredging. In the Philippines, ports are now government controlled but the Subic Bay former US naval base could become a private port facility, while in Malaysia the container terminal at Port Kelang was privatised in 1986 and in 1992 a private company was created to control the entire port apart from port safety and vessel traffic system. Even pilotage has been vested in the Kelang Management Company and overall there have been cost savings and improved productivity (*Fairplay*, 16/9/1993).

Arguably it was in some South American ports that efficiency was at its lowest and cost at its highest and where there was the most need for change (*Lloyd's List*, 1/10/1992). While Buenos Aires is still under the government and several other ports come under provincial control a degree of autonomy in the French style has been granted to Rosario, Santa Fe, Bahia Blanca and Quequen. A new port law of 1992 opened the way for some private facilities and grain traders such as Cargill, Dreyfus and Ramallo have now invested in grain terminals with marked improvements in handling rates and cost reductions. In Brazil about 75 per cent of all tonnage, mainly bulk commodities (e.g. iron ore at Tuburão), is handled at private terminals but since 1993 such terminals have been allowed to handle goods other than their own – Tuburão has made some grain shipments and Cargill is opening up its grain terminals to other users.

In Colombia there have always been private bulk terminals but since 1993 all ports have been 'commercialised' under regional companies. Chile also has long had private terminals (coal, wood chips) but has been working towards regional devolution of general cargo facilities and in 1993 at Puerto Ventanas, Quintero Bay, the first private general cargo terminal was opened. In Peru, where ports are controlled by the navy, labour and union monopoly has been swept away and at Ilo port management has been put out to tender. In Mexico, Puertos Mexicanos is to be liquidated and

273

control of ports will pass to local government, which it is intended will have a landlord status, with terminals operated on concession by the private sector. In 1995 the concessions for the operating of container terminals at Veracruz and Manzanillo were granted respectively to Mexican/Philippine and Mexican/US consortia.

Port development is investment intensive, especially for large bulk terminals and container facilities, and the capital and other resources (technical expertise, management, training) have often been in the form of aid (Robinson, 1989) from the World Bank or on a government-to-government basis and not always with proper consideration of the particular needs of the port, the danger of foreign influence or even control over developments and no real concern for the success of the enterprise. With privatisation the avenues for finance will undoubtedly be diversified, with large industrial, shipping and trading companies likely to be dominant – the dependency may still be there, possibly in a more blatant commercial form, but with the participants themselves having a vested interest in the success of the port there may well be a greater concern for efficiency and productivity than has hitherto been the case. While the government will no longer carry any burden for possible loss making facilities it may be desirable to create a body to produce a national port plan and ensure that individual port developments are within the framework this provides.

# 8

# DEVELOPING THE
# MARITIME CAPABILITY

Some Developing Countries have a long and significant maritime trading tradition. Over several millenia, India, with its 5,400-km coastline, has had long-distance seaborne trade westwards with the Red Sea, Egypt and East African coast and eastwards with Indonesia and Indo-China. The dhow trade of the Indian Ocean was well established long before the European navigators found their way into the area in the fifteenth century and Chinese archaeology has established the existence of a shipbuilding technology to sophisticated standards as far back as the Ch'in dynasty of 200 BC. Chinese sea-going junks 'incorporated maritime innovations like the watertight compartment, balanced rudder and spoon-shaped stern at a time when Westerners were still hugging coastlines in open longboats' (Maitland, 1981) and may well have known the Cape of Good Hope 1,500 years before the Europeans. Over a similar time span the Malayo-Polynesian peoples developed a strong seafaring tradition and as in the case of the Chinese junks, the great variety of 'prahus', the Indonesian generic ship type, suggests very long periods of evolution (Horridge, 1986) and long-distance, regular trade.

However, the arrival of Europeans in the Americas, Caribbean, Indian Ocean and Pacific was certainly to transform trading patterns and set in train the emergence of the global economy in which all countries are now enmeshed. Many of the traditional trades continued but the Europeans gradually assumed the controlling influence and Europe became the main focus of the transformed, long-distance trading patterns and the base for technological innovation e.g. the iron ship, steam propulsion. The new maritime technology facilitated industrialisation and colonialism and produced the core–periphery and dependency relationships which still characterise the global economy.

It has been suggested that maritime connections and ports have a particular significance for Developing Countries (Nagorski, 1971) and a number of reasons may be advanced for this. In many such countries the domestic markets are very small, external trade makes up a high proportion of their GDP and their resource endowments make them natural competitors rather than trading partners. They are invariably dependent on a limited range of primary exports with low value in relation to weight for which transport costs may make up a high proportion of the delivered price and therefore assume critical importance in determining their ability to compete. Most of the consumer and capital goods required by their expanding economies have to be imported because they are not produced locally. For Africa as a whole, as little as 4 per cent of the total trade is cross-border and the rest is maritime, while for Colombia the comparable figure is 5 per cent.

The external trade is therefore dominated by western, overseas partners who, over time, have come to control the shipping services. Yet transport costs, in which shipping is often a substantial component, may account for between 30 and 50 per cent of all foreign-exchange earnings and this provides a strong incentive to develop a domestic shipping industry. However, in order to reduce the cost of handling their trade they may have to adopt the newer, more productive technology (Chapter 7) and the ships that go with this – and they are having to do this at a time when the technology itself is changing rapidly, with the real danger that costly mistakes will be made that cannot be corrected easily or quickly (Hilling, 1980b). A new ship will have a life of 20 to 25 years while port facilities could last for a century.

The particular geography of some developing regions and countries (e.g. archipelago: Philippines, Indonesia, Caribbean, Pacific; peninsula/island: Malaysia) makes shipping indispensable for servicing domestic trade and may be an important factor in the maintenance of national cohesion (Brookfield, 1980). Indonesia's emphasis on extending and improving its domestic inter-regional ferry services is a case in point (Rutz, 1987).

While there can be no doubting the economic significance and therefore broader development relationships of shipping there is a real sense in which it is extra-territorial and this combines with historical factors already noted to explain why in many Developing

Countries there has been a tendency to leave shipping to the vagaries of external forces. Although not necessarily different in technology it is convenient to distinguish between what may for convenience be called the 'informal' and 'formal' sectors of shipping.

## THE 'INFORMAL' SHIPPING SECTOR

The long-standing maritime tradition of some Developing Countries is reflected in a range of present-day shipping activities and while much of this is domestic or coastal in character there are situations where it takes the form of international, deep-sea links.

In their study of the 'country' boats of Bangladesh, Jansen *et al.*, (1989) identify sea-going craft of traditional designs, usually of wooden construction with sail and/or motor propulsion. In addition to coastal links between Chittagong and Dhaka some venture southwards to ports in Myanmar. The Malayo-Polynesian peoples have a tradition of long-distance voyaging (Horridge, 1986) and many of the Indonesian island communities (Bajan, Buginese, Makassarese, Madurese) are still builders of sea-going craft involved in inter-island and international trade. Larger vessels, such as the *nade* of Sumatra and Riau, trade widely in timber and charcoal in the whole Malaysia-Indonesia region and the Buginese *pinisi* are found far from their home ports.

Long-distance Chinese trading junks, now mainly motorised, can be seen over the whole South China Seas and the continued importance of the dhow in the Gulf region is confirmed by 1990s plans by Dubai to extend the dhow facilities at Hamriya – dhow trade already amounts to over one million tonnes a year and is increasing at an annual eight per cent. While much of the trade is within the Gulf itself there is still long-distance trade with East Africa, Pakistan and India. In the Caribbean some inter-island links are still provided by sailing schooners.

What these operations have in common is the small size of the individual enterprises and an almost complete dependence on local enterprise, management, capital, skills and labour. While the amount of capital may initially be relatively modest there is considerable scope for modernisation to improve speed, reliability, capacity and range. The *pinisi* of Indonesia were, until recently, purely sailing vessels but have now been motorised (Hughes, 1981), as have the long-distance junks and dhows. The larger trading dhows, 100 to 500 dwt, are being fitted with marine diesels and long-range fuel tanks – and this in hulls of traditional design and timber construction, with rudimentary crew accommodation and often the simplest navigating equipment. A modern dhow could mean an investment of $700,000 and it is only where there is a reasonable expectation of plenty of cargo, high utilisation and quick port turn-round that this is justified – the considerable high-value consumer goods trade between Dubai and Iranian ports provides just such conditions and supplements a base load of locally produced fruit, vegetables and livestock (*Dock and Harbour Authority*, April-May, 1993).

The next step in the modernisation process is likely to be the introduction of sea-going vessels of modern design, construction and propulsion. Initially, these may well be smaller, short-sea vessels from European waters, which find second owners in Developing Countries. The operations are still at the scale of local initiative but may involve larger amounts of capital and possibly more sophisticated support organisation on shore (office, agents, bunkering) – they are becoming more formal.

The trade carried by the less formal sector of shipping is of two main types. In many areas the most important part comprises locally produced and traded goods – charcoal from Indonesia to Singapore and Malaysia, salt in Bangladesh, sawn timber in Indonesia. However, many of these island and coastal regions are at the extreme periphery of the global trading and shipping system and local shipping provides the essential means of concentrating individually small quantities of locally produced goods for export (e.g. copra from scattered Pacific islands) or distributing the whole range of imports. The dhow traffic from Dubai epitomises this with televisions, electrical goods, household equipment and motor cycles providing typical cargo. The trade generated by such regions is often very dispersed, volumes small and directionally imbalanced, not the conditions likely to justify attention by larger, main-line shipping companies (Couper, 1979; Dunbar-Nobes, 1984).

There is evidence to suggest that in recent years there has been an increasing marginalisation of such regions as larger shipping lines, which once made calls, now concentrate their activity at a smaller number of main ports in order to reduce operating costs. In some places even these main ports are under threat (Dunbar-Nobes, 1984). In this situation, the small, locally based ship

*Plate 8.1* The *pinisi* of Indonesia are vital for inter-island trade but like large dhows and junks have now been motorised for greater range, speed and reliability.

operator is likely to provide the only feeder services and critical external links on which the peripheral region depends. In the Pacific region there is thought to be a lack of functional complementarity between the ocean and local carriers and too little attention by governments to the provision of suitable feeder services (Couper, 1979; Brookfield, 1980) and the possible need to subsidise links that would not be viable.

It has been suggested that in the Pacific region the organisation of domestic shipping is of five main types (Drew, 1989) and there are certainly other areas where a similar classification would seem appropriate. The individual, private-sector, *indigenous operators* tend to use smaller, more traditional craft and often have their origins in the local communities with which they therefore have strong links. Where they are successful these operators may move up-market by increasing the number, size or sophistication of the vessels they use. There are also examples of *expatriate individuals* operating their own vessels, usually as the owner-master of smaller, coaster type. *Private-sector corporate* organisations often provide larger vessels which may offer dedicated services for a particular company's movement of raw materials or trade goods and they may be either under local or foreign ownership. Although of declining importance, there are still areas where *church ownership* has allowed missions to provide valuable transport services over and above their own needs. Finally, there are services provided by *government*, particularly in situations where the locations served and the nature of the traffic make economic viability uncertain but where, nevertheless, links are thought necessary.

Even where government-controlled, these operations are likely to be under-capitalised, using cheaper, older vessels for which maintenance may well be inadequate, breakdowns frequent and spares difficult to come by or not easily financed. The problems of distance and navigation may mean that the vessels used are larger than the trade really demands and this adds to the financial problems (Drew, 1989). As the local economies expand so the provision of shipping will become less precarious.

Globally, a large number of island and isolated coastal communities depend for their survival on small-scale shipping which ranges in technology from the purely traditional, through the smaller modern (although possibly second-hand and old) to up-to-date modern vessels. It would be impossible to calculate the overall contribution of such shipping in terms of the tonnages moved but it must be considerable and is of critical significance for the communities concerned and a main determinant of their present level of economic development and potential for growth – for many it is the only form of long-distance trading connection.

## THE FORMAL SECTOR

While there is considerable scope for local initiative in the provision of domestic shipping it is more difficult for the Developing Country to become involved in formal-sector provision of deep-sea international services. There are a number of reasons why this should be so, foremost of which are their colonial history and the role of overseas trade dependent on political control and economic forces generated by industrial and shipping enterprises in the colonising countries. The concentration of the trade of the former British West African territories in the hands of just a few British trading houses and the almost complete domination of the shipping until recently by the Elder Dempster company has been chronicled by Davies (1973). He argues that the shipping lines, defending their monopoly control through the conference system (see below), were able to reap monopoly gains and provide capital resources for the further expansion of trade in the region – a highly competitive shipping industry would have provided less profit and reduced surplus for reinvestment.

Shipping was the essential element which allowed the spread of the capitalist system and the expansion of markets which were basic to regional specialisation and division of labour and hence economic development. Broadly speaking, two main types of shipping operation can be identified – 'liner' and 'tramp'.

Liner shipping provides regular, often time-tabled, services between set ports at fixed freight rates and the ships convey the whole range of general cargo, sometimes with limited space for refrigerated cargo or liquids (e.g. vegetable oils, latex). This type of shipping requires expenditure on overheads (advertising, agents, supporting technical and managerial services in the ports served) and is vulnerable in the face of price competition (Stopford, 1988). It was to eliminate such competition that the liner companies serving particular routes invariably joined forces in the form of a 'conference' (see below).

In contrast to the liner system is 'tramp' shipping, although the term is less used now than formerly. This system operates by inducement in the open market (the 'spot' or 'charter' market) and is often one ship for one cargo (Stopford, 1988) – ship loads of dry bulk commodities (such as coal, ores, grain, forest products, fertiliser) or liquids (crude oil, petroleum products). It is very much a case of 'following the trade', with charters arranged through brokers at an exchange such as the Baltic Exchange in London. This is a highly volatile market and the freight rates will reflect the day-to-day relationship between demand and supply. This involves considerable risk but does not require the shipowner to have a costly, shore-based organisation. It is a more suitable form of ownership for smaller enterprises – the large number of Greek, family shipowning companies demonstrates this.

In most former colonial territories there was neither the local capital nor the entrepreneurs and commercial support at the right scale to make shipowning feasible and for the most part it was therefore controlled by large firms based in the mother countries. Even on political independence, earlier in South America and more recently in Africa and Asia, the underlying economic determinants persisted and many of the economies remained essentially 'colonial' in structure Thus, even by the 1950s the longer-standing, independent countries such as Argentina, Brazil and Chile had still made virtually no impact on global shipping.

## THE INTERNATIONALISATION OF SHIPPING

In the post-1945 period the global empires were gradually broken up and large numbers of new independent states were created. There was a new stimulus and more positive approach to development and economic planning and the expansion of national shipping was seen as a critical element in progress towards greater economic independence. There was mounting concern at the perceived discrepancy between the trade generated by Developing Countries and their involvement in the shipping of that trade – in 1967 Developing Countries accounted for 63 per cent of world trade loaded and 19 per cent of the goods unloaded yet controlled only 7.4 per cent of world shipping.

Transport costs may account for up to 40 per cent of balance of payments deficits for some of the lowest-income countries and the shipping is operated not as a service to them but for the profit of the shipping company. Indeed, it has been suggested (Yeats, 1981) that the Developing Countries systematically paid higher than average shipping costs – a differential that has been widened by the adoption of more productive, lower-cost cargo-handling methods for the trades between advanced economies (e.g. containerisation). The cost of shipping between Developing Countries and between them and the industrialised countries is higher than that between advanced economies. This has been advanced as a factor impeding trade between Developing Countries in South America (Yeats, 1981). Industrialisation and local processing of goods for export can have the effect of putting the goods into higher tariff categories and this has to be countered by the adoption of cheaper cargo-handling technologies.

The adverse relationship between transport costs and balance of payments has fuelled the grievances of the Developing Countries and stimulated their desire to develop their own fleets. In this way they hoped to influence global shipping strategies and reduce their transport costs and dependency on others for the movement of their trade. However, it can be argued that the colonial shipping system actually provided a degree of protection

*Plate 8.2* Small, traditional sailing craft still provide long distance international transport for cargoes such as salt and charcoal (here at Malacca).

in trade and shipping and the newly independent country was suddenly in a more vulnerable position and at the mercy of the conferences and charter market – freedom could be costly (Gold, 1981).

## FLAG AND REGISTER

It is traditional practice for a ship to fly the flag of the country to which it 'belongs', but this begs questions as to what constitutes belonging. What conditions determine the country to which a shipping company or its ship belongs and how genuine should those links be? The registration of the ship under a particular flag gives that ship its nationality and the ship becomes subject to the laws of that country. The flag state has legal jurisdiction over the operation of the ship, whether in home waters or elsewhere.

It has always been the case that shipowners sought to fly the flag that was most convenient to them at a particular point in time and changed flag as conditions changed. In the sixteenth and seventeenth centuries English ships were at times put under the French or Spanish flags to allow them unhindered trading in the West Indies and during the 'prohibition' period in America in the 1920s some shipowners switched to the newly created (1922) Panamanian registry. From such moves emerged the concept of the 'flag of convenience' (FOC), a term in common use by the 1950s (Tolofari, 1989) although the alternative 'open registry' is now favoured. Some registers are 'open' in the sense that they do not demand any genuine link between the ship, its owning company and the flag state, e.g. they do not require that beneficial ownership is in the hands of nationals, that the principal place of business is in the state, that the crew be nationals, that the company be taxed. The open registry is providing a commercial, money-earning service – it is selling its flag/nationality to shipowners who find its use economically or operationally convenient in comparison with their 'home' flag. The FOC state does not necessarily have any interest in imposing its national sovereignty and control over the ships flying its flag but it provides a source of income for countries which may lack other exploitable resources – 'the Madeira register was . . . another source of income for the weak economy' (*Fairplay*, 27/1/1994).

The ease of access to FOCs – registration with a consul abroad in some cases – accounts for much of their popularity. Moreover,

Table 8.1   World fleet by type of economy (percentages of total)

|                              | 1968 | 1981 | 1991 |
|------------------------------|------|------|------|
| OECD                         | 59.7 | 51.0 | 33.7 |
| Former USSR/Eastern Europe   | 8.0  | 7.2  | 8.1  |
| Developing Asia[a]           | —    | 4.2  | 7.0  |
| China                        | —    | 1.8  | 3.3  |
| Other Developing Countries   | 6.6  | 10.0 | 11.3 |
| Open registries[b]           | 16.1 | 25.0 | 35.6 |
| Other[c]                     | 9.6  | 0.7  | 1.0  |

Source: OECD, Maritime Transport, annual
Notes: a Hong Kong, Taiwan, Singapore, South Korea, Malaysia
   b Antigua and Barbuda, Bahamas, Bermuda, Cayman Islands, Cyprus, Gibraltar, Honduras, Lebanon, Liberia, Malta, Oman, Mauritius, Panama, St Vincent and Grenadines, Vanuatu
   c Cuba, Falkland Islands, Indonesia, Israel, North Korea, South Africa, Vietnam

a simple registration fee based on tonnage, lack of manning restrictions, lack of local taxation and lack of resources and competence to exercise proper compliance with international regulations make such flags attractive to shipowners in traditional maritime countries where taxation regimes, crewing requirements and conditions of employment and safety regulations serve to increase costs. The FOC can also be a haven for the unscrupulous shipowner.

The first open registry was Panama in 1922 but since 1949 many other countries have provided this service – Honduras, Liberia, Cyprus, Somalia, Bahamas, Vanuatu, Oman, St Vincent and the Grenadines and Malta to name a few main flags. From a mere 4.1 million GRT (5 per cent of the world fleet) in 1950 the FOC fleet had grown to 155 million GRT or 36 per cent of the world tonnage in 1991. It is clear from Table 8.1 that overall, the FOCs have grown very rapidly and are now the largest single category of registration – ahead of the combined fleets of the OECD countries. The world's leading flags are now the principal FOC states of Liberia and Panama (Table 8.2) with Cyprus and Bahamas having rapid recent growth. The size of these fleets represents neither their national shipowning nor their trading requirements – the beneficial ownership of the ships resides mainly with companies in the OECD countries and the United States, Japan and main European maritime states in particular.

*Table 8.2*  World fleet by principal flags

|  | 1970 | 1981 | 1991 | Tanker | Dry bulk | General |
|---|---|---|---|---|---|---|
|  | *(millions GRT)* | | | *(Percentages for each flag)* | | |
| Liberia (O/R) | 33.3 | 74.9 | 52.4 | 51 | 28 | 13 |
| Panama (O/R) | 5.6 | 27.7 | 44.9 | 31 | 30 | 33 |
| Bermuda (O/R) | 0.7 | 0.5 | 3.0 | 65 | 7 | 5 |
| Cyprus (O/R) | 1.1 | 1.8 | 20.3 | 30 | 15 | 18 |
| Bahamas (O/R) | — | — | 17.5 | 50 | 28 | 14 |
| Malta (O/R) | — | — | 6.9 | 35 | 40 | 22 |
| Hong Kong | 0.6 | 2.6 | 5.9 | 13 | 67 | 16 |
| Singapore | 0.4 | 6.8 | 8.5 | 42 | 25 | 30 |
| China | 0.9 | 7.6 | 14.3 | 13 | 36 | 45 |
| India | 2.4 | 6.0 | 6.5 | 28 | 48 | 17 |
| South Korea | 0.8 | 5.1 | 7.8 | 7 | 57 | 26 |
| Philippines | 0.9 | 2.5 | 8.7 | 4 | 73 | 20 |
| Taiwan | 1.3 | 1.9 | 5.9 | 11 | 46 | 40 |
| Indonesia | 0.6 | 1.7 | 2.3 | 25 | 6 | 47 |
| Saudi Arabia | — | 3.1 | 1.3 | 43 | — | 42 |
| Iran | — | 1.2 | 4.6 | 64 | 23 | 9 |
| Argentina | 1.3 | 2.3 | 1.7 | 32 | 21 | 31 |
| Brazil | 1.7 | 5.1 | 5.9 | 33 | 46 | 12 |
| | | | | | | |
| Main O/Rs | 40.7 | 105.3 | 155.2 | | | |
| OECD | 143.9 | 214.7 | 146.9 | | | |
| World | 224.3 | 420.8 | 436.0 | | | |

*Source:* OECD, *Maritime Transport*, annual
*Note:* (O/R) – Open registry

## ACCESS TO THE SHIPPING MARKET

In theory, anyone with the capital and necessary inclination can purchase a ship and operate it on the open market which is mainly concerned with shipping of bulk cargoes – dry and liquid. The organisation is relatively simple and the scale can vary from the single ship to large fleet operations. The freedom of access and large number of firms operating makes this as close to perfect competition as shipping comes and it is the type of market in which a Developing Country would not be disadvantaged.

However, as shown in Table 8.3, these are just the markets in which the FOCs are strongly represented and other Developing Countries relatively weakly. In the very markets in which they generate so much trade (oil, mineral ores, coal) the Developing Countries are having to compete with the ships of advanced economies and multinational corporations operating under the shelter of FOC. Not surprisingly, the FOCs are not popular with other Developing Countries who see them as an impediment to the expansion of their shipowning aspirations, providing as they do a means whereby ships owned in high cost countries are able to reduce their costs and compete more effectively. This also explains why many Developing Countries are opening their flags to outsiders.

## THE LINER CONFERENCE SYSTEM

In contrast with the open, charter market which characterises the bulk trades, most general, break-bulk cargo (the whole range of small consignments of capital and consumer goods and bagged cargo such as cocoa and coffee) was carried mainly by the liner companies with their complex organisation and greater investment in ships and shore-side services. Many of the liner companies became very large and powerful (e.g. Elder Dempster in the West African trade, Blue Funnel in Asia) but characteristically the lines serving a particular route grouped themselves into what became known as 'conferences' (Deakin and Seaward, 1973; Janssen and Shneerson, 1987; Stopford, 1988). The first such conference was on the UK–Calcutta route in 1875 and was a response by the liner companies to intense competition, overcapacity, low freight rates and inability to cover full costs. In the

*Table 8.3* World fleet by ship type, 1991 (percentages of total)

|  | Oil tankers | Dry bulk | Container/ RO-RO | General cargo |
|---|---|---|---|---|
| OECD | 35.1 | 28.1 | 44.2/52.7 | 18.7 |
| Former USSR and Eastern Europe | 3.7 | 4.1 | 3.0/23.6 | 13.7 |
| Developing Asia[a] | 4.2 | 10.8 | 11.8/1.2 | negl. |
| China | 1.3 | 4.5 | 3.5/1.4 | 10.0 |
| Other Developing | 11.9 | 17.1 | 11.6/8.1 | 25.4 |
| Open registries[b] | 43.6 | 35.2 | 23.2/12.8 | 30.0 |
| Other[c] | 0.1 | 0.2 | 2.7/0.2 | 2.1 |
|  | 100 | 100 | 100/100 | 100 |

*Source:* Calculated from OECD, *Maritime Transport,* annual
*Notes:* a, b, c – as for Table 8.1

conference the shipowners pooled their resources, regulated the sailings, shared the traffic, fixed the freight rates and shared the revenue. In the so-called 'closed' conferences, which is most of them in fact, access is restricted and membership limited and loyalty rebates or inducements are given to shippers who use only conference vessels. The conference was in effect a cartel to rule out price competition. The number of conferences peaked at about 400 and is now about 250 (Bridges, 1992).

Apart from the perceived need for such cartels by the shipowners it could be argued that the shippers derive some benefit through regularity of sailings (often fixed-day departures), a high level of reliability, good quality, specialised tonnage suited to particular trades (e.g heavy-lift derricks for logs in West Africa) operated by companies with long experience of those trades and freight rates that were published in advance. If the freight rates were high it reflected the quality of service and cost of providing it – the ship sailed, full or not.

Critics argued that the rates were fixed too high and to suit the shipowner rather than the shipper and could in fact be altered without consultation and with little warning (e.g. the imposition of a congestion surcharge). The semi-monopoly control is undesirable in principle, encourages bureaucratic control, facilitates political interference and reduces incentives to innovate and improve services. There is no general agreement on the balance between advantages and disadvantages and indeed it probably varied between conferences and from time to time. Nevertheless, there was a widespread feeling in Developing Countries that the conferences acted to their disadvantage and some studies certainly pointed (*West Africa*, 25/7/1994) to the high price of conference shipping and lack of response to shipper needs. In 1991 the Continental-West Africa Conference was fined by the European Commission for operating under monopolistic conditions.

There was a growing concern that it was the conferences which held back the development of their shipping although there was undoubtedly ambivalence on this. The Developing Countries were attracted to liner shipping with its higher earning potential but dissatisfied with the conferences which were thought to exclude them from an area of decision making which was vital to their development. Payments to foreign shipowners were seen as a major drain on foreign-exchange reserves. Yet it has to be said that when they created their own shipping lines they immediately sought to join the conferences and were invariably admitted promptly.

The criticism of the liner conferences was taken up by the Group of 77, the Developing Countries, at UNCTAD and at Santiago in 1972 a draft Code of Conduct for Liner Conferences was tabled, was adopted by UNCTAD in 1974 and came into force in 1984. The main objectives were fourfold – to encourage fleet development in Developing Countries, to increase their earnings from shipping, to improve their balance of payments and to regulate the behaviour of the liner conferences. The Code makes provision for the easier admission of new lines, the institutionalisation of consultation arrangements (e.g. shippers' councils), puts constraints on the increasing of freight rates and creates machinery to settle disputes.

The aspect of the Code which attracted most attention and certainly most dispute was the 40–40–20 Market Sharing Formula (MSF). The object was to ensure fair participation by all parties by stipulating that in the trade between any two countries each should be entitled to carry 40 per cent of the trade with the remaining 20 per cent available to third parties. The Code was self-regulated and therefore in practice unenforceable and by the time it came into force in 1984 much liner shipping had become containerised, there was wide adoption of uniform Freight All Kinds (FOK) rates which altered pricing policies, there had been a rapid growth in shipping by 'outsiders', carriers operating outside the conference system, which had seriously dented conference monopolies and some Developing Countries already carried 40 per cent of their trade – where they did not there was no clear way by which they could be brought up to this level. Arguably the Code was too late and in 1992 was in effect abandoned.

However, it has been shown that even had it been fully implemented the redistributional effect would not have been that intended (Wergeland, 1985). There may well have been some redistribution between traditional flag states but few Developing Countries would have gained benefit and some would have lost (Table 8.4) – the FOCs dramatically so. If the potential redistribution is recalculated on the assumption that all known cargo reservations are taken into account some developing regions benefit rather more, but the impact is certainly not dramatic and hardly justifies all the effort that went into creating the Code of Conduct.

It has already been suggested that at least in theory it is easier

*Table 8.4*  Potential redistribution of liner vessels through 40-40-20 market sharing

|  | 1 | 2 | 3 | 4 |
|---|---|---|---|---|
| EEC | 16.1 | 25.9 | 10 | −168 |
| Scandinavia/Denmark | 3.9 | 2.3 | 92 | −39 |
| North America | 30.0 | 13.7 | 298 | −89 |
| Japan | 6.8 | 18.9 | 100 | 44 |
| Australia/New Zealand | 6.6 | 3.9 | 73 | 6 |
| Other OECD countries | 0.6 | 0.8 | −44 | −19 |
| Central America/Caribbean[a] | 4.0 | 1.7 | 53 | 71 |
| East Coast South America | 2.9 | 2.6 | −30 | 40 |
| Other South America | 2.9 | 1.7 | −15 | 23 |
| South Africa | 1.9 | 1.3 | 18 | −3 |
| Med. Africa/ Middle East | 1.2 | 7.6 | −19 | −51 |
| North & West Africa | 2.0 | 2.2 | 6 | 42 |
| East Coast Africa[b] | 0.8 | 0.7 | 5 | 10 |
| South Asia | 1.7 | 2.6 | −101 | −20 |
| South East Asia[c] | 8.7 | 2.9 | 106 | 132 |
| Far East (excl. Japan) | 3.9 | 6.3 | 87 | 125 |
| Oceania (excl. Australia/New Z.) | 0.3 | 0.2 | 0 | 3 |
| Centrally Planned Economies | 3.6 | 3.5 | −304 | 52 |
| Open Registry Countries | 1.7 | 1.4 | −336 | −159 |

*Source:* Wergeland (1985)
*Notes:* 1. Percentage of world tonne-miles, imports
   2. Percentage of world tonne-miles, exports
   3. Redistribution on strict 40-40-20 shares
   4. Redistribution after restrictions taken into account
   a – Excluding Panama, b – Excluding Somalia, c – Excluding Singapore

for Developing Countries to enter the bulk shipping charter market and with this in mind there has been some pressure for a bulk shipping MSF of 50–50 and the abolition of the FOCs which account for most of such shipping. This is being strenuously opposed by both the FOCs and the traditional shipping nations.

Many countries adopted policies which suited their particular circumstances even where in practice that meant ignoring the Code. Initially, South Korea argued that all its trade should be carried in Korean vessels except where such were not available or where a Korean shipowner waived the right to carry the cargo (Jin Goo Jim, 1993). This was later relaxed to liner cargo only and for certain designated, strategic bulk cargoes (e.g. oil, iron ore, grain, coal) and is in process of further relaxation in recognition of the fact that reservation policies may indeed be harmful – efficiency of resource allocation is reduced, excessive investment in shipping may be encouraged and the country may have no comparative advantage. With low liner rates the opportunity costs of capital may not be covered, shipping is expanded for non-shipping purposes, easy access to reserved cargoes could reduce incentive for high standards in shipping, shipping is highly competitive and there is no point in having a fleet that is not effectively used. There may be economic advantage in buying in low cost shipping services from elsewhere. All Developing Countries intent on expanding their fleets should perhaps keep these cautions in mind.

In West Africa the liner companies created by Ghana (Black Star Line), Nigeria (Nigerian National Line) and other countries became fully involved in the appropriate conferences and in 1994, the China Ocean Shipping Company (COSCO) abandoned a go-it-alone policy and initiated moves to participate in conferences.

## FLEET DEVELOPMENT

While progress may not have been at the pace nor in the direction envisaged by the Developing Countries the composition of the world fleet has changed dramatically in the last 30 years, with decline of the traditional maritime nations, represented in OECD, and the massive growth of the FOCs being the main features. While the FOCs are for the most part Developing Countries, it is not the case that most Developing Countries offer open registers. The Developing Countries' fleet, excluding FOCs, grew from 16.2 to 22.6 per cent of the total between 1968 and 1991 but much of this was accounted for by China and Pacific Rim countries best classed as NICs (Table 8.2). India has seen considerable fleet expansion as have some of the Middle East oil exporters and in South America Brazil stands out. However, there has been little growth in most other South American and African countries and the fleets of many of the latter have in fact declined – Ghana and Nigeria are obvious examples.

Developing Countries appear destined to become the operators of older ships. Having to replace the ships bought in the first phase of expansion some 20 years ago but with insufficient capital to invest in new ships they can only buy second-hand and often older

*Plate 8.3*   In Indonesia, as in other areas, more modern small craft
provide scope for local enterprise and investment.

vessels of less efficient design and propulsion for which maintenance and operating costs may be higher than average. Late in 1994
the state-owned Pakistan National Shipping Corporation was given
approval for a fleet renewal programme with loan guarantees by the
government but will be able to afford only second-hand tonnage.
There is a danger in this situation that if the necessary maintenance
is not carried out the old vessels become unsafe vessels.

Much of the tonnage under Developing Country flags is of the
general cargo type and they have only 12 per cent of the world
tanker and 17 per cent of the dry-bulk tonnage (Table 8.3). Only
those Developing Countries with large bulk imports are likely to
derive real benefit from involvement in tramp shipping in bulk
trades. However, increasing proportions of general cargo are now
containerised and this needs examining in more detail.

## THE CONTAINER TRADES

The Developing Countries are not well represented (Table 8.3) in
the high-technology general cargo trades (containers, RO/RO)

which are dominated by the traditional maritime nations and several NICs, the Taiwanese Evergreen Line and the South Korean Hanjin Line being the obvious examples. A report of 1994 pointed to an actual decline in the poorer countries' share of container shipping at a time when the total tonnage was expanding rapidly, as was their share of other tonnage (United Nations, 1994). South Africa apart, in Africa only three, very small container ships were under local control.

Containers are ideally suited for high-value manufactured goods and in the trades between industrialised countries have reached their full potential. Because it is often imbalanced or unsuitable in quantitative and qualitative terms much of the trade of Developing Countries does not favour containerisation and certainly not the use of fully cellular container ships – some container capacity on conventional ships may be the most that is needed. Both Ghana's Black Star and Nigeria's National Shipping lines had such 'combo' tonnage and in 1992 the Iran Shipping Line ordered 12 general cargo ships with substantial container capacity. Less efficient ports and hinterland links and the greater bargaining power of big shippers in negotiation with the conferences for large cargo volumes were also cited as disadvantages for the Developing Country trying to build up container tonnage (United Nations, 1994). As industrialisation increases so too will the demand for container shipping, as demonstrated by Taiwan, Hong Kong, Singapore and Malaysia.

World-wide weakening of liner conferences and the clear economies of scale associated with container shipping have resulted in the emergence of global mega-carriers with restricted port itineraries and this provides scope for servicing feeder routes to the main hub ports (Frankel, 1987; Turner, 1991). The smaller vessels on such feeder routes will not require the great investment of the global services, will be less competitive but do create the dependence on external hubs – much of India's container trade is by transhipment through Singapore and but for political problems Colombo might have captured more of this trade (Turner, 1991). On the feeder routes there is scope for locally owned shipping possibly in partnership arrangements with main-route shipping companies who seek maximum control over the feeders to their services. In Asia some of the feeder operators have expanded outwards to become genuine intra-regional carriers.

With massive port congestion in the ports of many Developing

Countries in the 1970s, RO/RO was seen as an appropriate technology to speed up ship turn-round in port (Chapter 7). The initiative for this came largely from the shipowners of developed countries and the Developing Countries have been little involved in providing tonnage. An exception has been the sophisticated RO/RO tonnage of the Brazilian Transroll company. In 1993, the state-owned Shipping Corporation of India announced plans for the country's first coastal RO/RO services between Bombay and Cochin and between Gujurat and Bombay and there has been increasing use of RO/RO for inter-island links in the Philippines.

## PUBLIC SHIPPING ENTERPRISE

Given the risks associated with the shipping industry it is reasonable to ask whether or not it is wise for Developing Countries to become involved and if the answer is yes, the form that it might best take (Smith, 1989). Except at the simplest, informal, domestic level, involvement in shipping has been mainly by state enterprises and there are few Developing Countries that have not had a state-owned shipping company, and most still do. Whether for doctrinaire, political reasons, rational, strategic considerations or a feeling that private enterprise was inadequate for the purpose, state-controlled shipping became the norm.

An extreme example of this is Ecuador, where the national line, Transportes Navieros Equatorianos (Transnave) is owned by the navy and run by an admiral – 'the navy is in shipowning to do things that the private sector can't or won't – transport is vital to Ecuador's foreign trade, we need liner shipping, and we have to support the skills and capacity to control transport' (*Fairplay*, 19/5/1994). These are sentiments undoubtedly shared by the governments of many Developing Countries – if trade was to be carried under the national flag it would have to be a state-owned line.

Established about 25 years ago, Transnave is comparable with many such state-owned lines. It faces intense competition, it has ageing tonnage and needs to move into container carrying. The Venezuelan state shipping line (CAVN) has six old conventional ships which are costly to run, debts have accumulated and the staff is far larger than needed for its operations. The Nigerian National Line in 1993 was in default on premium payments for insurance cover and in recent years there have been frequent reports of its vessels being arrested for non-payment of debts. In 1995 the company went into liquidation. By 1994, Ghana's Black Star Line from a peak of 12 vessels was down to three, dating from the early 1980s, with heavy fuel consumption, high maintenance costs and large, 41-man crews. In 1992, the Peruvian state line went into liquidation. Brazil's shipping had in 1978 carried 30 per cent of the country's trade but by 1993 this was down to 3.2 per cent.

Many of the state-owned lines suffer from lack of investment, unsuitable tonnage for changing demands of trade, high operating costs and poor management and the high-cost service they provided was propped up by various forms of reservation of cargo for national flag ships (Ademuni-Odeke, 1988). Many South American countries, in particular, have come in for criticism for operating extreme forms of cargo reservation but this can be an expensive and inefficient way of protecting national flag shipping. Countries such as Venezuela, Ecuador, Sri Lanka and India have varying degrees of deregulation, although this can open up the state-owned lines to competition from domestic, private or foreign shipping with lower cost structures. So burdensome have some of the state-owned lines become that there have been moves to sell them off. Argentina, Bangladesh, Brazil, Ecuador, Peru, Philippines, Puerto Rico and Turkey have all been engaged in divestment but often the lines prove not to be attractive to private enterprise. China's 30-year old state line, COSCO, has been split into more autonomous operating units, joint ventures have been entered into and in 1994 the country's first private deep-sea shipping line, Dafeng, was created.

Not all state-owned shipping has been characterised by inefficiency and malaise. The United Arab Shipping Company (UASC), a multi-state enterprise dating from 1974, now has a diversified fleet of some 50 vessels including container tonnage. Indonesia's Petramina has shipping as an integral part of its oil and gas production and processing and in 1994 had 172 tankers of all sizes, many of which were involved in the particular, domestic distribution problems of a large archipelago country, but arguably constrained by the inflexibility that comes with state control. Transportacion Maritima Mexicana is one of the largest lines in South America, with impressive results in recent years, but since 1993 has effectively been operated as a private company. India has the advantage of a large domestic trade and the state-owned Shipping Corporation of India (SCI) has become one of the world's largest fleets although it has undoubedly been hindered by lack of foreign exchange. Between 40 and 80 of the line's vessels are due for scrapping but most recent purchases have been in the second-hand market and there is little container capacity.

Created in the 1970s, the Malaysian International Shipping Corporation (MISC) was for long synonymous with the country's deep-sea shipping and by 1994 had a fleet of 45 vessels, including 15 container ships. Malaysia provides a good example of a country in which positive nation building and economic development strategies have encouraged investment in all areas, including shipping, and MISC is a substantial and successful enterprise. However, together with privately owned companies, it still carries only 15 per cent of the country's trade and the cost to the economy of foreign flag shipping is set to double by the year 2000 – hence the strong incentive to encourage national flag shipping.

## PRIVATE SHIPPING COMPANIES

State-owned lines provided the pioneer, formal-sector, deep-sea shipping for most Developing Countries and private enterprise was mainly restricted to the informal sector described above. The move from this small-scale involvement to the level of deep-sea shipping is far more demanding of capital and managerial skills and few entrepreneurs were able to make the transition. It has been suggested that everywhere, the emergence of the professional ship owner was a comparatively recent phenomenon (Ville, 1987) and historically, shipowning was usually based on reinvestment of profits from other business activities. Not surprisingly, private shipowning will be best developed in those countries in which local entrepreneurial groups have emerged with a capacity for capital accumulation and investment and where economic growth has been most pronounced.

For obvious reasons the NICs are likely to be in the forefront and a supreme example is provided by Taiwan's Evergreen Line which has grown, from one conventional ship in 1968, to become the world's largest container ship company, with pioneer round-the-world services and diversification into feeder services, shore terminals, land transport, liner agency, hotels and in 1989 an airline. As Taiwan prospered so did Evergreen (Table 8.5).

In Colombia, bulk shipping apart, most of the private shipping is controlled by members of the powerful, coffee growers' association and in the early 1990s Brazil's privately owned deep-sea fleet

*Table 8.5*   Development of Taiwanese Evergreen Line

| | | |
|---|---|---|
| 1968 | Evergreen Marine Corporation created to serve Far East – US West coast trades | Purchased *Central Trust*, 10,000 dwt second-hand, conventional 'tween deck vessel |
| 1974 | Started full container service, Far East–New York | Acquired 646 TEU container ship *Ever Spring* Three sister ships added |
| 1979 | Started Far East–Europe container service | *Ever Spring* class ships jumboised to 866 TEU 1,214 TEU 'V' Class introduced |
| 1984 | Introduced first round-the-world container service | 2,728 TEU 'G' Class introduced 3,428 TEU 'GX' Class introduced 4,229 TEU 'R' Class introduced 4,900 TEU Post-Panamax ships introduced |
| 1992 | 51 vessels | |

numbered some 150 vessels, although this was declining because of government insistence on new ships being built in Brazil, the high cost of such ships, high interest rates and crewing regulations which added to operating costs. Peru's Maritime Association has over 20 companies but most of the ships are registered under cheaper FOCs.

The Scindia Steam Navigation Company was India's best known private shipping line, having come into existence soon after independence, but when it had financial problems in the late 1980s it was taken over by the government. There are now 15 or so Indian companies operating in deep-sea trades. The Great Eastern Shipping Company was started by Vasant Sheth in 1948 when he bought the first Indian flag tanker. Sheth saw the potential in shipping India's considerable exports of coal and mineral ores and the Bombay company has built up a fleet of over 40 vessels and in 1995 took delivery of seven ships valued at $60 million – some new and some second-hand. Century Shipping is a recently created branch of a large cement/chemicals/pulp and textiles combine and operates bulk carriers, tankers, chemical tankers and log carriers. There are now six companies operating out of

Madras with Pearl Ships, a joint venture between a local and a Norwegian company, involved in dry-bulk shipping.

In 1992 Chinese control of shipping was reformed and this allowed easier access to the market for local and foreign investment and a number of new shipping companies were started. Initially these were involved in domestic trades but the Jujian Shipping Company and Dafeng Company now operate deep-sea. Thailand has a Maritime Promotion Commission and in 1993 private companies owned 242 vessels divided almost equally between dry cargo and tankers – ships and companies are mainly small and only three (Regional Container Lines, Jutha Shipping and Precious Shipping) had raised funds on the stock market. In 1995, Precious Shipping bought two mid-1980s handy-size bulk carriers (bringing the fleet to 33 vessels) to ship rice needed to balance poor harvests. Filipino companies tend to 'flag out' but an Overseas Shipping Development Act of 1992 gave incentives for the purchase, rather than chartering, of vessels; as the world's leading provider of seafarers the country needs as many ships as it can get – irrespective of the flags that they fly (*Fairplay*, 5/8/1993).

Concerned with the rapid expansion of the economy and the need to provide adequate links between its thousands of islands, the Indonesian government in 1988 opened up the country's shipping to foreign competition and made it very difficult for local companies to compete in domestic trades and almost impossible in deep-sea trades. Joint ventures provide a possible solution to this problem and companies may have to diversify into stevedoring, warehousing or freight forwarding in order to survive (*Lloyd's List*, 27/4/1993).

## SHIPBUILDING AND FINANCE

Shipping companies of Developing Countries can acquire their ships either on the sale-and-purchase (second-hand) market or as new buildings. In the former, price will depend on the age and condition of the vessel, freight rates, inflation and market predictions (Stopford, 1988) but many smaller companies may have to go for older tonnage. The life of a ship would normally be 20 years but many older vessels are still sailing – South American coastal waters have been called a 'dump for old ships' (*Lloyd's List*, 1/8/1995). New ships, while more expensive, are more likely to attract financial aid through shipyard subsidies and grants (Branch, 1982) either from domestic or foreign governments (e.g. export credits). In Developing Countries few shipowners are able to finance ship purchase entirely from their own funds and grants and loans invariably play an important part – the availability of concessional finance is usually a critical factor and the shortage of such funds and especially of foreign exchange has often been a main constraint.

India's Shipping Credit and Investment Company (SCICI), Brazil's Marine Fund and the Development Bank of the Philippines are examples of state support for shipping and the World Bank, Asian Development Bank and private financial institutions have also become involved. In 1992 India made it easier for its shipping companies to buy and sell ships, approach foreign lenders for funds and use their foreign exchange earnings for ship finance. In the Caribbean it is thought that companies involved in deep-sea trades have obtained financial assistance from banks more readily than smaller operators on inter-island trades. It has been suggested that an international credit company should be created, similar to India's SCICI and possibly structured along the lines of the World Bank and sponsored by the UN's International Maritime Organisation (*Fairplay*, 19/9/1991). There is growing evidence that some of the problems of finance are being tackled by joint ventures and while these should be approached with caution there is no reason why both parties should not gain (Abhyankar and Bijwadia, 1994).

With respect to new buildings few of the Developing Countries have shipbuilding capacity for anything but smaller vessels of an unsophisticated nature (Table 8.6). The African capacity for larger

*Table 8.6*   Ships on order by building country, June 1994[a]

| | | |
|---|---|---|
| OECD | 753 | (47%) |
| Former USSR/Eastern Europe | 345 | (21%) |
| Developing Asian economies[b] | 299 | (19%) |
| China | 111 | (7%) |
| Other Developing Countries | | |
|     Asian | 42 | (3%) |
|     South American | 30 | (2%) |
|     African | 9 | (0.5%) |
| World | 1,607 | (100) |

*Source*: Calculated from *Fairplay Newbuildings*, July 1994
*Notes*:  a  Excluding passenger and some miscellaneous craft
      b  Malaysia, South Korea, Taiwan

deep-sea traders is entirely in Egypt and that in South America is accounted for by Brazil and Argentina. Furthermore, and following from the problems of finance, Developing Countries account for a very small proportion of ships on order (Table 8.7). The FOCs apart, and most of their new buildings are not beneficially owned in Developing Countries, Africa and South America account for a mere 2.5 and Asia only 3.0 per cent of all orders. NICs apart, India and Indonesia account for most of the Asian orders and only China at 5.0 per cent is making any real progress. With new building at this level and OECD countries accounting for more than 50 per cent of the buildings in all types of vessel (80 per cent of RO/ROs) the shipowning gap is widening at an alarming rate.

*Table 8.7*  Vessels on order by country of owner, June 1994

|  | Total | General | Container | Tanker | Bulk | RO/RO | Other |
|---|---|---|---|---|---|---|---|
| OECD | 980 | 194 | 136 | 256 | 232 | 40 | 122 |
|  | (54) | (55) | (52) | (55) | (55) | (80) | (50) |
| Former USSR/ | 197 | 90 | 11 | 54 | 19 | 7 | 16 |
| Eastern Europe | (11) | (26) | (4) | (11) | (5) | (14) | (7) |
| Developing Asian | 274 | 11 | 68 | 58 | 106 | 2 | 29 |
| economies[a] | (15) | (3) | (27) | (12) | (25) | — | (12) |
| China | 88 | 12 | 9 | 26 | 26 | — | 15 |
|  | (5) | (3) | (4) | (5) | (6) | — | (6) |
| Other Developing | 139 | 25 | 15 | 67 | 9 | — | 23 |
| Countries | (7) | (7) | (6) | (14) | (2) | — | (9) |
|   Africa | 11 | 0 | 0 | 2 | 2 | 0 | 7 |
|   South America | 40 | 3 | 15 | 11 | 5 | 1 | 3 |
|  | (2) | — | (6) | (2) | (1) | — | (1) |
|   Asia | 54 | 2 | 0 | 26 | 3 | 0 | 10 |
|  | (3) | — | — | (5) | — | — | (4) |
| Open Registry | 38 | 1 | 11 | 5 | 8 | 0 | 13 |
|  | (2) | — | (4) | (1) | (2) | — | (5) |
| Unknown | 90 | 17 | 4 | 24 | 19 | 1 | 25 |
| World | 1,806 | 350 | 254 | 490 | 419 | 50 | 243 |
|  | (100) | (19) | (14) | (27) | (23) | (3) | (13) |

*Source:* Calculated from *Fairplay Newbuildings*, July 1994
*Notes:*  (a) Hong Kong, Malaysia, Singapore, South Korea, Taiwan, Thailand.
      Figures in parentheses below numbers of vessels are the percentages of world total for each type of vessel.

## CREWING AND EXPLOITATION OF SEAFARERS

In August 1995 a large Danish shipping company announced that 100 of its Danish junior officers were to be replaced by Indians and Filipinos – quite simply, they are cheaper. While the Developing Countries' share of world shipping may be small their share of seafarers is very large (Brooks, 1985) and it has been claimed that 'when a ship sinks, there is strong possibility that Filipino seamen will be among the dead and missing' (*Fairplay*, 30/6/1994), reflecting the fact that 14,000 vessels, roughly half those trading, have Filipino crews – in all about 130,000 seamen with another 80,000 available. Regrettably what this reflects is a cost-reducing option.

There is a long-standing tradition that ships of traditional maritime nations employed seamen from overseas territories – P&O used Indian, Blue Funnel Chinese and Elder Dempster West African crews. There was a clear link between the crews employed and trading region and invariably the seafarers served in lower-grade deck, engine-room and catering jobs. In recession many companies have cut back on officer training and for a country such as Britain the number is now inadequate to replace natural loss, with a possible knock-on effect for the many shore jobs (pilots, harbour masters, shipping company shore staff, etc.) which depend on sea-going experience to a high level (McConville, 1995) and may in future have to be fillled by non-nationals. Also, wage rates in the traditional maritime countries have risen steeply, forcing shipowners to seek cheaper alternatives. With crew costs accounting for about 50 per cent of operating costs (Stopford, 1989) there are strong incentives to do this. A 1993 survey by the International Shipping Federation showed large variations in monthly salary:

|  | Highest | Lowest |
|---|---|---|
| Chief Officer | $14,000 | $700 |
| Able Seaman | $7,000 | $300 |

Undoubtedly, many are earning less than these figures suggest and there have been cases of unions opposing wage increases for fear of losing jobs to even cheaper sources – Filipinos may be cheap at the moment but there are plenty of seamen from the Maldives, Myanmar, Indonesia, Sri Lanka and, possibly, China ready to take their places.

Some think that for the most part life at sea is attractive only to those unable to find suitable shore employment! If this is so, then in the early stages of economic development there are certainly 'push' factors encouraging many to go to sea and break out of the poverty cycle – 16,000 supposedly applied for 80 places at one Indian sea school (*Lloyd's List*, 4/12/1991) and in the Philippines there are now well over 250 agencies in the crewing business, 72 of which are members of the Filipino Association for Maritime Employment (FAME) which is concerned with achieving acceptable standards. In some countries – Myanmar is the best example – the government provides the crewing agency and earns as much as $150 million a year from the placement of its nationals. As wage opportunities on shore increase so this demand to go to sea may decline but in many Developing Countries we are far from this situation. South Korea once provided a large number of seamen but far less are now available and seafaring has declined in social status. Hong Kong and Singapore, for their part, provide very few seamen.

With 70 to 80 per cent of marine accidents caused by human factors and growing concern for ship safety and the dangers of environmental pollution there is a related concern for the quality of crews. In the Philippines the proliferation of sea schools has not been with due regard for standards, has often been without arrangements for built-in sea experience and with evidence of widespread, fraudulent practices. Pass rates of 96 to 100 per cent for the different grades of deck and engine-room officer were thought to be too good to be true and the introduction, with the support of FAME, of computerised preparation of questions and marking produced more convincing rates of 8 to 45 per cent (*Lloyd's List*, 29/4/1993).

The Monaco-based V Ships management company, which employs large numbers of Filipinos, has set up its own training courses in conjunction with the Cavite Polytechnic and operates two of its vessels as training ships. Norwegian companies have joined forces with Greek, Hong Kong and Japanese interests to set up a training school in the Philippines.

India has a long tradition of providing seamen, especially for British ships, and its maritime training infrastructure, the skill of its seamen and the regulation of its certification system are highly regarded (*Lloyd's List*, 17/7/1992). However, a cut-back in the 1980s has left even India short of trained officers although union agreement to reduced manning levels in 1994 has helped. A feature of the 1995 revision of the International Maritime Organisation's Standards of Training, Certification and Watchkeeping will require far more regulation by national governments and quality assessment of training establishments.

Seafarers have traditionally been an exploited group (Hugill, 1967) and the conditions under which they were expected to live and the pay they received were often very poor. Undoubtedly, far too many shipowners take advantage of the low-cost crews and have no concern for the welfare of those they employ.

Cases have been reported (*Trade Winds*, 27/8/1992) of crewmen forced to work 200 hours overtime a month at rates between $0.50 and $1.00 an hour and of seamen living on a bowl of rice and one dollar a day. There is widespread payment at below internationally agreed rates and, all too frequently, late and even non-payment. Unions and national governments connive in this. For a small country, such as the Maldives, the 3,000 nationals employed in foreign ships bring valuable foreign exchange and those recruited have to sign a Maldive Government contract which forbids union membership and which guarantees shipowners against additional claims which might be made by the International Transport Workers Federation (ITF) (*Trade Winds*, 22/1/1993). It is thought that Maldive seamen may have been jailed at home for causing trouble in their employment. Myanmar seamen are recruited through a government agency and Amnesty International was asked to investigate the disappearance of 11 Myanmar seamen who had forced a payment for overtime while their ship was in an Australian port and then refused to be flown home.

It seems clear that large numbers of seamen are not being paid or treated in an acceptable way but so important is the direct or indirect benefit to their countries that their governments refuse to support them for fear of making them unemployable. The adequacy of existing conventions, lack of regulation to ensure observance and general consideration of wages, conditions and manning are to be high on the agenda for the 1996 Maritime Session of the International Labour Organisation.

## FLAG, REGULATION AND SAFETY

The International Maritime Organisation's Safety of Life at Sea (SOLAS) convention excludes from full compliance 'ethnic' craft and small vessels involved in domestic trades. Some 300 vessels trade out of the Miami River to all parts of the Caribbean and many of these are small, owner-skippered, probably not classed by a reputable ship-classification society (e.g. Lloyd's), not likely to be properly insured and often with faulty equipment (*Fairplay*, 8/7/1993). Overloading at remoter, less public wharves is normal. This description would be applicable to a large number of vessels trading in many other parts of the world which technically come under SOLAS and for which FOCs provide nationality and the regulatory authority.

Many factors contribute to ship safety and casualties (routes, cargoes, loading and unloading, design, standard of construction, age, maintenance) but there is mounting evidence that flag and crew may be of over-riding significance but, as implied above, these are related, with flag often selected to provide less stringent regulation of crewing. Three aspects of crewing are critical (*Lloyd's List*, 26/11/1993) – skill and experience level, communication between crew members of different language and culture and problems of officer leadership – all typically problems on many FOC ships.

In the seven years to 1992 the loss rate for ships registered in Panama, Cyprus, Bahamas, Malta, Vanuatu and Liberia was twice the world average and it seems clear that under these flags, with Liberia a possible exception, the flag regulations are less rigid and through lack of resources, ability or inclination the international regulations are enforced weakly or not at all. With many flag states not providing checks on ship and crew standards the problem has had to be tackled from a different direction – the port state. With major European countries and America in the lead, but other areas following, emphasis is now put on the inspection of ships entering ports and this port-state control is now resulting in regularly published 'hit-lists' of flags and companies whose ships have to be detained on account of deficiencies of a physical or human kind (Table 8.8). Certain FOCs frequently appear on the defaulter lists: for example, Cyprus, Panama, Malta and St Vincent and the Grenadines would claim to be victims of targeting. That this is unreasonably discriminatory is not proved. However, there is a sense in which many of the Developing Countries are allowing under their flags ships which are sub-standard and some of which display a lethal combination of old age, poor maintenance and unsatisfactory crews. Hitherto, port-state inspections have concentrated on vessel fabric, equipment and documentation but there is a growing demand for more attention to crew related issues of competence and conditions. This would be justifiable on the

*Table 8.8*   Port State Control ship detentions

a) United Kingdom Port State inspections/detentions – year to May 1995

|  | *Number of detentions* | *Percentage of ships inspected* |
|---|---|---|
| Malta (0/R) | 24 | 27.3 |
| Cyprus (O/R) | 23 | 13.4 |
| Russia | 23 | 10.4 |
| Panama (O/R) | 16 | 20.3 |
| Honduras (O/R) | 9 | 40.9 |
| St. Vincent & Grenadines (O/R) | 7 | 17.5 |
| Bulgaria | 7 | 38.9 |
| Turkey | 6 | 55.4 |
| Lithuania | 6 | 18.8 |

b) Detentions as percentage of inspections in Australian ports, 1993

| | |
|---|---|
| St Vincent and the Grenadines | 33 |
| Honduras | 25 |
| Malta | 19 |
| Indonesia | 11 |
| Netherlands Antilles | 10 |
| China | 9 |
| Turkey | 9 |
| Egypt | 8 |
| India | 8 |
| Philippines | 8 |
| Malaysia | 6 |
| Russia/Ukraine | 5 |
| Cyprus | 4 |
| South Korea | 4 |
| Panama | 4 |
| Norway | 3 |
| Hong Kong | 2 |
| Japan | 2 |
| Liberia | 2 |
| Greece | 1 |
| Singapore | 1 |

*Sources:* a) Dept of Transport
b) Australian Port State Inspection Authority
*Note:* (O/R) – Open registry

grounds both of ship safety, the main concern of port-state inspectors, but also as a way of countering the more obvious forms of crew exploitation.

All the evidence suggests that in trying to build up their fleets, either by providing open registries or simply national flags, many Developing Countries have contributed to a lowering of standards in shipping. While in the short term this may be advantageous for the income it generates, in the long term it could be counter productive and Liberia and Cyprus have certainly responded by tightening the controls – others may well follow, faced with the prospect that their ships will be detained on port-state inspection. The flags of some Developing Countries are now being used by unscrupulous shipowners and for the exploitation of nationals of Developing Countries and it must be in the long-term interest of these countries to prevent this. In building up their own fleets, special aid may be necessary to ensure that they are not forced to rely on old, second-hand tonnage for which maintenance costs will be prohibitively high and that they can provide training to acceptable international standards, both in their shore establishments and also at sea. This may best be achieved by way of partnership arrangements of the type which the Philippines is developing but as economic development proceeds so local initiatives are likely to become more important – South Korea providing the obvious example.

# 9

# SOME THEMES FOR THE FUTURE

There can be no doubting the all-pervasive influence of transport and in particular its critical role in the process of development in the broadest sense – economic, social and political. Transport is both the cause and an effect of development and the precise nature of the interrelationship will certainly continue to provoke debate. However, a better understanding of the problems involved could lead to developments in transport which produce benefit for the whole populations concerned and a widening rather than concentration of advantage and an emphasis on equity rather than polarisation. There has been no attempt in this volume to provide definitive answers but rather to develop ideas and facts which will inform discussion and decision making.

In the advanced economies where existing transport networks are highly developed, extensive and well integrated, particular new developments, while they may ease the movement of people or goods, rarely have any significant impact on the broad patterns of economic and social life. This is not the case in most Developing Countries where there is still vast scope for transport projects to influence, even determine, the direction and pace of the whole development process. Any improvements to transport should increase the range of and/or the capacity for spatial interaction and in large parts of the less-developed world the present level of transport provision is so inadequate that any improvements hold out considerable scope for change in other sectors of activity. It is, therefore, the more important that decisions with respect to transport are made only after the fullest consideration of the society's development goals and the possible ways in which any new transport could affect the attainment of those goals. There are in the Developing Countries all too many examples of transport projects

*Plate 9.1* Transport has always provided a challenge to local initiative – as demonstrated by this wheel-steered donkey cart in rural Jamaica – it is important that such enterprise is not stifled.

which have not been thought through with sufficient rigour, which have had little or no direct development impact and which, based as they frequently are on borrowed money, have done little more than add to the debt burden. This they cannot afford.

## CHOICE OF TECHNOLOGY

It is not enough to think simply in terms of transport in general. This book reflects the author's interest in the technology of transport; decision making with respect to transport projects is certainly about choice of mode and for each mode the type and level of technology – what in Chapter 1 was called the *'what'* factor. Arguably, too many transport planners have placed too much emphasis on the hardware of transport and failed to recognise that a particular technology will be appropriate only to the extent that it is properly related to a particular point in time – the *'when'* factor – and also to the broad spectrum of human and physical geography – what may conveniently be termed the *'where'* factor.

All transport is place-specific and the local conditions are of vital importance – a point argued strongly by Wright (1992) and Dimitriou (1992) with respect to urban areas but equally applicable to passenger transport in rural areas and freight transport generally. It is important to have the right transport in the right place at the right time. Where transport projects have failed to stimulate development it can usually be explained in terms of inadequate consideration of the interrelationship between the 'what', 'when' and 'where' factors. The World Bank and other aid agencies are becoming more stringent in their selection procedures to ensure that money is well spent (Armstrong-Wright, 1986, 1993; Robinson, 1989).

It is also worth asking *who* is making the decisions? Too many decisions are handed down from the top and it is not unusual for 'the top', whether urban-based politician or foreign consultant, to be insufficiently in touch with demand at the grassroots level – and with theorists (Pawson, 1979) now favouring the demand led model it is important that provision is related to the actual demand and not some possibly erroneous perception of it on the part of a remote decision maker. In defence of foreign consultants it has been argued that it is in the follow-up, rather than the execution, of the projects that the faults are to be found (Farahmand-Razavi, 1994). This clearly brings us to ask *why* decisions are arrived at. There is the case of an African port which was apparently located for no better reason than that it was the home of the Minister of Finance and there is at least the suspicion that big, rather than most suitable, may be all too beautiful for consultants and engineering contractors seeking business.

The years during which many Developing Countries have become independent politically have also seen revolution in transport technology with significant innovation in air transport and the advent of containerisation and the very large ship as main elements. While it would be incautious to predict no further innovations in transport in the next few decades (double-jumbo jets? high-speed ocean ships? electric cars?) it seems likely that there will be a period of consolidation with the further spread of tried technology rather than dramatic new developments. This will certainly be welcome to the Developing Countries for whom decision making with respect to basic infrastructure has not been easy while the technology available is changing rapidly.

In recent years there have been rapid advances in information technology and in some respects this could be an alternative to physical movement. Electronic data interchange is a normal way of moving documents related to freight transport. Increasingly the new technology will be finding its way into Developing Countries but could well be a factor in the further marginalisation of large

*Plate 9.2*   A World Bank financed road in Gabon has provided access to large forest resources – but is this the right transport, in the right place and for the right development?

areas where, for a long time to come, far more basic transport provision is the real priority.

## THE INTERNATIONALISATION OF TRANSPORT

When dealing with air and maritime transport (Chapters 4, 7 and 8) the increasing internationalisation was noted and there is a sense in which even at the local level the consolidation of goods for export and the distribution of imports is part of an international chain of transport operations often involving several modes and numerous providers.

While the technological revolution in transport has had far-reaching implications for spatial interaction there have been related changes in thinking about the organisation of transport systems (Hayuth, 1987). The new technology has made possible and spawned a whole new set of ideas and terminology – 'intermodalism', 'through-transport', 'one-stop shipping', 'physical distribution', 'product channel management' – all reflecting a new integration, often at the global scale, of transport operations. Where previously the elements in the transport chain were quite separate in control and organisation there is now a desire to bring them together in large, functionally unified organisations. Many shipowners are no longer just that – a company like P&O is now an intermodal operator with interests in warehousing and the consolidation and distribution of freight by its own land transport. The object is to control the operation from origin to destination.

Only the largest operators have the resources to spread themselves in this way and an inevitable consequence has been the emergence of the large transnational transport operator (P&O, Maersk, Evergreen, SeaLand) and consortia of smaller companies to achieve critical mass with respect to resources and to derive the benefits of economies of scale. CONcentration of activity allows DOMination of the market – the CONDOM principle in operation.

Another consequence of this trend is the emergence of hierarchical structures in the global transport system with hub-and-spoke patterns of movement. For the Developing Countries there are in these trends the possible dangers of further marginalisation, reduced control over transport operations and increased dependency on others for basic transport services – the reverse of everything for which they strive. However, this may not be all loss – land-locked countries are beginning to benefit from through freight rates and bills of lading and this will facilitate their international trade.

For reasons already described (Chapter 4) air transport has become highly regulated on an international basis and as a result of growing concern over ship standards, crewing and safety there is evidence that shipping is moving, albeit slowly, in a similar direction. This is coming about through the International Maritime Organisation's conventions with respect to Safety of Life at Sea (SOLAS) and Marine Pollution (MARPOL), the adoption of more stringent criteria by some ship classification societies and insurers and by growing concern for quality assessment in the shipping business. Enforcement of IMO conventions by flag states is very patchy but there is increasing pressure to conform to internationally acceptable standards through port state inspection and detention of ships found sub-standard, either structurally or with respect to crew competence. A number of Developing Countries will need to upgrade their tonnage, some may be forced out of the shipping business and others will have to cease being havens for sub-standard vessels.

At the level of domestic air and sea services the imposition of international standards is difficult, even impossible. Nevertheless, the all too frequent and sometimes heavy loss of life on, for example, heavily overloaded inter-island ferries in the Philippines, suggests that there is the need for acceptable standards to percolate downwards.

## MOVEMENT OF PEOPLE

In most Developing Countries the present provision for passenger transport is woefully inadequate yet as development proceeds, rapidly in some places but depressingly slowly in others, there will be continued and sometimes rapid growth in demand in both rural and urban areas. Everywhere the rate of urbanisation is alarmingly high and shows no signs of slowing down in the near future. Over the decade of the 1990s the urban population of Africa will increase by 4.7 per cent, of Asia by 3.0 per cent and Latin America by 2.6 per cent – for India this means that the urban population is growing at 600,000 every month (Armstrong-Wright, 1993). There have been many reminders that this could be the critical area in transport decision making and provision over the years to come (Owen, 1987; Wright, 1992). The present grossly unsafe overloading of many buses, trains and ferries may now be necessary but will become increasingly unacceptable as incomes and expectations rise.

Evidence from elsewhere suggests that rising incomes will certainly lead to increased demand for personal transport – as they rise to middle levels this could be for motor cycles and further rise will mean a growing desire for cars. Either way, it will add to already congested roads and lead to declining efficiency and reduced ability of public transport to meet growing demand from lower income groups. There are strong grounds for arguing that restraint on car use, nowhere popular, should be imposed before too many have come to take the convenience of the car for granted.

Certainly for a long time to come the growing volume of motorised personal transport will be competing for road space with a very wide range of traditional vehicles, both for passengers and freight, the numbers of which will also be expanding. A mix of vehicles of differing speeds tends to get reduced to the lowest common denominator, network capacity is lowered and congestion mounts. The financial resources to deal with this problem by massive road building programmes are not available in most Developing Countries and it is open to debate whether or not this is a solution which should be encouraged. Likewise, as indicated in Chapter 6, there are probably relatively few cities in which the conditions favour and resources are available to support rail mass transits as a solution – in most cases they would not be that. Developing Countries will have to adopt solutions which are less demanding on scarce resources of all kinds and which instead place an emphasis on effective management of the available infrastructure to ensure maximum capacity utilisation.

Armstrong-Wright (1993) suggests that such management may take three forms. Transport derives from a demand which is in large measure a function of land-use patterns and more attention must be given to *land use management* and planning so as to reduce the demand for movement or facilitate that movement. Too many people live too far from their place of work but as economies expand and employment is created, new developments should be located to reduce the gap.

While it will not necessarily solve the basic problem, *traffic management* offers a relatively low-cost way of easing traffic movement. There is considerable evidence of ill discipline in the use of road space and vehicles (Joshi, 1981; Badejo, 1990) and an essential starting point should be a more effective separation of different types of vehicles (bus lanes, slow-vehicle exclusion zones, priority schemes, recommended routes) and of vehicles and other road users (pedestrians, street traders). On all too many routes there is a complete free-for-all and this must be replaced by regulated traffic and enforced discipline – not easily achieved unless corrupt practices are also eliminated.

There is then the possibility of *demand management* although few places anywhere have had much success in this area – Singapore may be something of an exception with its well-developed, zonal car-use pricing scheme but it is perhaps worth noting that it is also a city with a level of discipline and order found in few other places. Attempts to modify the temporal features of demand may not achieve much in cities characterised by high levels of informal activity and flexible working arrangements. Altering consumer preferences and enticing people away from their personal transport depends to a large extent on providing an acceptable level of public transport provision in terms of convenience, comfort and cost; and most Developing Countries still have far to go in this respect.

If the problems of the urban areas derive from excessive demand, in many rural areas the demand is at a low level and widely dispersed. Yet some 65–70 per cent of the population of the Developing Countries is in these areas and if they are not to be further marginalised in the development process they must be provided with levels of transport which facilitate the expansion of economic activity and incorporate them into the wider socio-economic structures. There can be no doubting the vast untapped potential of many of these areas (Beenhakker et al, 1987) and while in a number of places a start has been made on improving their access this has to be taken much further and much faster if they are not to fall further behind. For the decision makers in the Developing Countries, and those who may advise them, there will be critical priorities to be determined – hitherto there has arguably been undue emphasis on the urban population.

## THE ENVIRONMENTAL FACTOR

Transport is a critical factor in the creation of socio-economic environments, becomes a main element in the resultant landscapes and while initially its influence may overall be beneficial, as time goes on it can become a cause of environmental degradation. It has to be said that the environmental consideration has nowhere in the past been given priority in most transport decision making but concern over environmental degradation is now mounting, at a local and global level, and transport is seen as a main contributing factor. While the environment in many Developing Countries, especially in their urban areas, is far from acceptable, the idea that this is inevitable or that a 'good' environment is a luxury, which they should not strive for and cannot afford, has now to be dismissed.

If transport is a major contributor to pollution, mass car ownership is the principal component. While there seems to be little chance of reversing the trend in advanced economies there is time to avoid the extreme manifestations in many Developing Countries because car ownership levels are still very low and every possible attempt should be made to keep them so. While one could interpret this as an attempt to deny reasonable aspirations it should perhaps be seen as a means of avoiding the gross mistakes of the advanced countries.

The arguments must be in favour of seeking solutions which are environmentally sound and also sustainable; uncontrolled car-use growth is not that and planners should be avoiding solutions which

*Plate 9.3* Traditional Peruvian river craft 'improved' by the addition of 'long-tail' or outboard motor.

add to pollution of all kinds and which place increasing demands on energy which must be at increasing cost in financial and environmental terms. The transport planner can no longer afford to ignore the environmental implications of the strategies recommended and this will require careful consideration of the technologies available and the environmental impact of each will have to be carefully evaluated. The decision maker should have the best practicable environmental option firmly in mind when alternatives are being considered.

## PUBLIC VERSUS PRIVATE ENTERPRISE

At a number of points in this book the traditional emphasis in many Developing Countries on public, governmental involvement in transport provision has been noted. Further, the inadequacy and inefficiency of much of this transport and its inability to pay its way makes it a continuing drain on national resources. Given a scarcity of government money and the variety of calls upon it there is a widespread move away from the 'infrastructure' approach and a greater emphasis on 'commercial' funding – this is often associated with deregulation, liberalisation and privatisation. Ports, almost entirely government-financed until recently, are now, most of them, involving private capital in one way or another (Chapter 7).

For a Developing Country this may create special problems. Clearly, while the situation varies between countries, in many of them there is no large accumulation of private capital to draw on and with much transport infrastructure and hardware not coming cheap (Table 9.1), local private enterprise may be limited to the single lorry, taxi or small boat. A few larger manufacturing companies may be able to finance transport operations as an integral part of their total operations – brewery companies often do this. There may be an input from local communities in the form of labour for road construction but this will be limited to roads of a simpler type – however, this may well be the type of road that is most needed in many areas. Only the very largest mining companies are likely to get involved in railway construction or shipping.

Many Developing Countries are faced with severe debt problems in part a result of investment in transport which brings little or no immediate financial return. Many of them have had to adopt structural readjustment programmes which invariably involve

*Table 9.1*  Cost of transport provision – selected items, 1995

| Lorries | | Buses | |
|---|---|---|---|
| 3.5-tonne panel van | £16,400 | 8-seater | £18,000 |
| 5.6-tonne panel van | £26,700 | 29-seater | £39,000 |
| 13-tonne rigid | £33,000 | 49-seater | £107,250 |
| 17-tonne rigid | £35,400 | 53-seater | £131,500 |
| 38-tonne articulated | £63,000 | | |

*Ships*

| | |
|---|---|
| Second-hand 370-TEU container ship | £3.2 million |
| Second-hand 1,800-TEU container ship | £16.6 million |
| Second-hand 6,000 DWT tanker | £1.2 million |
| Second-hand (1976) 84,000 DWT tanker | £3.5 million |
| Second-hand 132,000 DWT tanker | £6.1 million |
| Second-hand (1974) 74,000 DWT bulk carrier | £4.0 million |
| Second-hand (1984) 28,000 DWT bulk carrier | £7.9 million |
| New 4,3000-TEU container ship | £70.0 million |

*Sources:* Lorries and buses – Croner, *Operational Costings for Transport Management* (1995) Ships – *Fairplay*

strict control of foreign exchange and the imposition of sound commercial practice on state enterprises (Filani, 1993). This makes it difficult to embark on new transport projects and requires existing services to pay their way. With government resources stretched to the limit and no great private capital available locally it follows that further external funding may have to be sought – this may be international, governmental-bilateral or private. The poor performance of so many aid-financed transport projects has led to official donors applying ever more restrictive appraisal criteria and in future only those projects with a high probability of success in economic, development-generation, poverty-alleviation or environmental terms are likely to be funded. It follows that transport developments will be on an ever more selective basis and will have to be evaluated in far more detail and prioritised more carefully than has hitherto been necessary.

Given the lack of local resources, even governments which have previously avoided doing so, now find themselves encouraging private, foreign investment in transport with the possible attendant dangers of loss of control, 'neo-colonialism' and increased dependency. Nevertheless, there is increasing foreign private financial involvement in ports, airports, airlines, shipping companies, road haulage and rail mass-transit systems. It may well be that joint enterprises provide the best way forward – either between local and foreign companies or between government and foreign private companies. It has been argued (Abhyankar and Bijwadia, 1994) that when carefully planned to ensure the equitable participation and benefit of both parties and with necessary safeguards against distortion there is no reason why joint ventures will not provide a satisfactory way of providing shipping services – and what is true for shipping is almost certainly true for other forms of transport provision.

# GUIDE TO FURTHER
# READING

The literature on the subject of transport and development is now vast and of necessity the bibliography for this book is highly selective. Even so, for the student approaching the subject for the first time, the list may seem daunting. Wilfred Owen's book *Strategy for Mobility* (1964) is in many ways the starting point for much later work and the ideas have been updated in his *Transportation and World Development* (1987), a useful global overview of the role of transport in all aspects of economic and social progress. Unwin's *Atlas of World Development* (Chichester: Wiley, 1994) illustrates global contrasts in levels of development across a range of phenomena. The volumes edited by Heraty in the *Developing World Transport* series provide numerous short, but interesting, case studies on the whole range of transport provision and policy making.

The Intermediate Technology Development Group, founded by the late Dr E.F. Schumacher in 1965, has a special interest in rural areas and appropriate transport development, and its ideas have been developed in a number of most valuable publications. Howe and Richards (*Rural Roads and Poverty Alleviation*, 1984) provide an excellent, comprehensive overview of the literature supported by case studies of Botswana, Egypt, India and Thailand, and Beenhakker et. al. (*Rural Transport Services*, 1987) draw on a wide range of sources to provide a guide for policy makers and planners. These, and other books, from Intermediate Technology Publications (Barwell *et. al.*, 1985; Edmonds and Howe, 1980) are firmly based on practical field experience and are a useful balance to academic theory.

Hadfield's *World Canals* (1986) is a comprehensive study of waterways past and present, including many examples from Developing Countries, and Intermediate Technology's *The Country Boats of Bangladesh* (Jansen *et. al.*, 1989) covers the natural and operational environments of inland shipping and by extension has much that is of relevance to other areas. Although concerned with North America rather than Developing Countries, Vance's lengthy, but readable, book *Capturing the Horizon* (1986) has much that is of interest on the theme of the impact of transport, and especially railways, on settlement, and the engineering aspects of railways are well covered in *Railroad Engineering* (Hay, 1982). Taneja's *Introduction to Civil Aviation* (1989) is a useful broad study of the international aviation industry with an emphasis on regional variations, and airline regulation, in the context of political, economic and social development, is well covered in Naveau's book *International Air Transport in a Changing World* (1989) and also in *A Geography of Air Transport* by B. Graham (Chichester: Wiley, 1995).

Based on wide experience in Latin America, Wright's book *Fast Wheels, Slow Traffic* (1992) is a provocative examination of problems endemic to urban transport and a valuable guide to transport strategies in cities. A wide range of papers on many aspects of urban transport, including rail mass transit, will be found in the proceedings of several conferences organised by the Institution of Civil Engineers (e.g. 1989).

Because shipping provides the means of linking resources and markets at the global scale it is an international and politicised industry. A comprehensive examination of the evolution and character of international shipping is provided in *Shipping and International Trade Relations* (Ademuni-Odeki, 1988) and the particular features and problems associated with open registry shipping are covered very adequately in *Open Registry Shipping* (Tolofari, 1988).

# BIBLIOGRAPHY

Abane, A. (1993) 'Modal choice for the journey to work among formal sector employees in Accra, Ghana', *Journal of Transport Geography*, 1(4): 219–24.

Abelson, P. (1995) 'Cost benefit analysis of proposed major rail development in Lagos, Nigeria', *Transport Reviews*, 15(3): 265–89.

Abhyankar, J. and Bijwadia, S. I. (eds) (1994) *Maritime Joint Ventures*, Paris, International Chamber of Commerce.

Ademuni-Odeke (1988) *Shipping in International Trade Relations*, Aldershot: Avebury.

Adler, H. A. (1971) *Economic Appraisal of Transport Projects*, Bloomington: Indiana University.

Akatsuka, Y., Asaeda, T. and Brooks, J. (1994) 'The outlook for research in Asian inland waterways transport', Brussels: Permanent International Association of Navigation Congresses, Special Bulletin, No 83/4.

Allport, R.J. (1991) 'The metro – determining its viability', in M. Heraty (ed.) *Developing World Transport*, London: Grosvenor Press International, 64–9.

Arlt, W.H. (1989) 'Port problems in developing world countries', in M. Heraty (ed.) *Developing World Transport*, London: Grosvenor Press International, 268–70.

Armstrong-Wright, A. (1986) *Urban Transit Systems – Guidelines for Examining Options*, Washington, The World Bank, Technical Paper No. 52.

—— (1991) 'Urban transport – World Bank policy', in M. Heraty (ed.) *Urban Transport in Developing Countries*, London: PTRC.

—— (1993) *Public Transport in Third World Cities*, London: HMSO.

Arrowsmith, B. (1995) 'Opportunities for port privatisation in China', *The Dock and Harbour Authority*, London: Foxlow, 59–64.

Badejo, B. A. (1990) *Private Operation of Public Bus Services: the Case of Metropolitan Lagos, Nigeria*, University of London, unpublished Ph.D. thesis.

Baker, J. (1974) 'Development of Ethiopia's road system', *Geography*, 59(2): 150–4.

Baldwin, M. (1980) *Waterways and Energy Saving*, London: National Waterways Transport Association, Paper No 2.

Barber, G. M. (1977) 'Sequencing highway network improvements – South Sulawesi', *Economic Geography*, 53(1): 55–69.

Bari, A. (1982) The design and evaluation of alternative inland waterway transport systems for developing countries, University of Newcastle-upon-Tyne, unpublished Ph.D. thesis.

Barlow, K. (1990) 'Air of change in Vanuatu', in M. Heraty (ed.) *Developing World Transport*, London: Grosvenor Press International, 263–6.

Barrett, I. M. D. (1986) 'Conventional bus operations in African cities', in Institution of Civil Engineers, *Moving People in Tomorrow's World*, London: Thomas Telford, 89–101.

—— (1988) *Urban Transport in West Africa*, Washington, World Bank, Technical Paper No 81.

Barwell, I., Edmonds, G.A., Howe, J.D.G.F. and DeVeen, J. (1985) *Rural Transport in Developing Countries*, London: Intermediate Technology Publications.

BBC (1984) *River Journeys*, London: British Broadcasting Authority, television series, book.

—— (1993) *Locomotion: The World the Railways Made*, London: British Broadcasting Authority, television series, video and book.

Beenhakker, H.L., Carapetis, S., Crowther, L. and Hertel, S. (1987) *Rural Transport Services*, London: Intermediate Technology Publications.

Bejakovic, D. (1970) 'The share of transport and communications in total investment', *Journal of Transport Economics and Policy*, IV(3): 337–43.

Best, T.D. (1966) 'Geographical factors and railroad gauges', Cornwallis, Oregon: *Yearbook Association Pacific Coast Geographers*.

Bird, J.H. (1963) *Major Seaports of the United Kingdom*, London: Hutchinson.

—— (1971) *Seaports and Seaport Terminals*, London: Hutchinson.

—— (1980) 'Seaports as a subset of gateways for regions', *Progress in Human Geography*, 4: 360–70.

—— (1983) 'Gateways: slow recognition but irreversible rise', *Tijdschrift voor Economische en Sociale Geografie*, 74: 196–202.

Blaikie, P., Cameron, J. and Seddon, D. (1977) *The Effect of Roads in West Central Nepal*, London: report for Overseas Development Administration.

Blair, J.A.S. (1978) 'The regional impact of a new highway in Sierra Leone', *African Environment*, June.

Bonayad, A. (1987) 'Cooperation among developing countries in world shipping', *Maritime Policy and Management*, 14, 267–9.

Bonney, R.S.P. (1964) *The Relationship Between Road Building and Economic and Social Development in Sabah*, Crowthorne: Road Research Laboratory, Paper LN/520.

Bonz, M. (1990) 'Criteria for the choice of light rail systems' in M. Heraty (ed.) *Developing World Transport*, London: Grosvenor Press International, 117–24.

Boyce, A.M., McDonald, M. and Pearce, M.J. (1988) *A Review of Geometric Design and Standards of Rural Roads in Developing Countries*, Crowthorne: Transport and Road Research Laboratory, Paper No. 94.

Branch, A.E. (1982) *Economics of Shipping Practice and Management*, London: Chapman & Hall.

Bridges, R.K. (ed.) (1992) *Directory of Freight Conferences,* London: Cromer.

British Waterways Board (1975) *Energy Conservation – the Inland Waterway Role,* London.

Britton, F.E.K. (1980) 'Transport: issues, priorities and options in the 1980s', in *Paratransit: Changing Perceptions of Public Transport,* Canberra: Australian National University.

Brookfield, H. (1975) *Interdependent Development,* London: Methuen.

—— (1980) 'The transport factor in island development' in R.T. Shand (ed.) *The Island States of the Pacific and Indian Oceans,* Canberra: Australian National University, Development Studies Centre, Monograph No. 23, 201–38.

—— (1984) 'Boxes, ports and places without ports' in B.S., Hoyle and D. Hilling (eds) *Seaport Systems and Spatial Change,* Chichester, Wiley, 61–77.

Brooks, Mary R. (1985) *Seafarers in the Asean Region,* Singapore: Institute of Southeast Asian Studies.

Browne, M. and Hilling, D. (1992) 'Bulk freight transport' in B.S. Hoyle and R. Knowles (eds) *Modern Transport Geography,* London: Belhaven, 179–98.

Brunn, S.D. and Williams, J.F. (eds) (1983)*Cities of the World,* London: Harper & Row.

Busby, R.H. (1989) 'Changing considerations and techniques in the selection and design of new railway systems' in M. Heraty (ed.) *Developing World Transport,* London: Grosvenor Press International, 184–8.

Button, K.J. and Pitfield, D.E. (1985) *International Railway Economics,* Aldershot: Gower.

Camara, P. and Banister, D. (1993) 'Spatial inequalities in the provision of public transport in Latin American cities', *Transport Reviews,* 13(4): 351–74.

Catling, D.T. (1986) 'Light rail concepts' in Institution of Civil Engineers, *Moving People in Tomorrow's World,* London: Thomas Telford.

Chaudhry, M.R. (1987) 'Proposal for inland navigation in Pakistan' CIT, Pakistan, *Proceedings of Seminar on Inland Water Transport,* Chartered Institute of Transport, Pakistan, Karachi, mimeo.

Chiu, T.N. and Chu, D.K.Y. (1984) 'Port developments in the People's Republic of China' in B.S. Hoyle and D. Hilling (eds) *Seaport Systems and Spatial Change,* Chichester: Wiley, 199–215.

CIT Pakistan (1987) *Proceedings of Seminar on Inland Water Transport,* Chartered Institute of Transport, Pakistan, Karachi, mimeo.

Clarke, C.W. (1957) 'Track loading fundamentals', *Railway Gazette,* 106, (1): 46–8; (ii): 157–63.

Cole, Viscount (1989) 'General aviation – the developing world's indispensable transport', in M. Heraty (ed.) *Developing World Transport,* London: Grosvenor Press International, 325–7.

Comtois, C. (1994) 'The evolution of containerisation in East Asia', *Maritime Policy and Management,* 21(3): 195–206.

Containerisation International (annual) London: Emap Response Publications.

Conway, C.J. (1990) 'Shipping operations: the developing world at sea', in

M. Heraty (ed.) *Developing World Transport*, London: Grosvenor Press International, 218–23.

Cook, J.H.G. (1989) 'Railways – construction and maintenance of the line', in M. Heraty (ed.) *Developing World Transport*, London: Grosvenor Press International, 188–90.

Cooley, C.H. (1894) *The Theory of Transportation*, Baltimore: The American Economic Association.

Coombe, D. and Mellor, A. (1986) 'The role of paratransit', in Institution of Civil Engineers, *Moving People in Tomorrow's World*, London: Thomas Telford: 103–16.

Couper, A.D. (1979) 'Ocean-going and domestic shipping problems for island countries', *Inter-Island Shipping Seminar Papers*, Singapore: Marintech.

Crawford, R.G. (1990) 'Market evolution and its impact on bus design', in M. Heraty (ed.) *Developing World Transport*, London: Grosvenor Press International, 98–102.

Cundill, M.A. and Byrne, H.M. (1982) 'A study of goods vehicle restraint in Bangkok', Crowthorne: Transport and Road Research Laboratory, Report No 733.

Dalvi, M.Q. (1986) 'The mobility problem of the Third World', in Institution of Civil Engineers, *Moving People in Tomorrow's World*, London: Thomas Telford: 151–67.

Dalvi, M.Q. and Saggar, R.K. (1979) 'Integrated transport policy and the role of inland water transport in national and regional development', ESCAP, *Planning the Development of Inland Waterways*, Bankok: ESCAP.

Danish Hydraulics (1988) Copenhagen: No 8.

Davies, P.N. (1973) *The Trade Makers: Elder Dempster in West Africa, 1852–1972*, London: George Allen & Unwin.

Deakin, B.M. and Seward, T. (1973) *Shipping Conferences*, Cambridge: Cambridge University Press.

Deplaix, J.M. (1989) 'Inland water transport – an economic and efficient mode', in M. Heraty (ed.) *Developing World Transport*, London: Grosvenor Press International, 291–3.

Deschamps, L. (1970) 'Interrelations between transport development and mass agricultural extension schemes', in D. McMaster (ed.) *Transport in Africa*, Edinburgh: Centre for African Studies University of Edinburgh.

Diallo, Y. (1994) 'Autonomous civil aviation administrations', Montreal: International Civil Aviation Organisation, *Journal*, Jan.–Feb.: 22–4.

Dick, H.W. (1981) 'The imperialism of urban public transport in Southeast Asia', Singapore: *Proceedings of the Pacific Science Association*, 4th Congress.

Dick, H.W. and Rimmer, P.J. (1980) 'Beyond the formal/informal sector dichotomy; towards an integrated alternative', *Pacific Viewpoint*, 21, 26–41.

Dickenson, J.P. (1978) 'Industrialisation in the Third World', in A.B. Mountjoy (ed.) *The Third World*, London: Macmillan, 93–101.

—— (1983) *A Geography of the Third World*, London: Routledge & Kegan Paul.

Dickinson, B. (1984) 'The development of the Nigerian port system:

crisis management in response to rapid economic change, 1970–1980', in B.S. Hoyle and D. Hilling (eds) *Seaport Systems and Spatial Change*, Chichester: Wiley, 161–77.

Dimitriou, Harry T. (1992) *A Developmental Approach to Urban Transport Planning; an Indonesian Example,* Aldershot: Avebury.

Dodgson, J.S. (1985) 'A survey of recent developments in the measurement of rail total factor productivity', in K.J. Button and D.E. Pitfield (eds) *International Railway Economics*, Aldershot: Gower, 13–48.

Doganis, R. (1985) *Flying Off Course,* London: Allen & Unwin.

Drew, O. (1991) 'Shipping in the Pacific island states', in M. Heraty (ed.) *Developing World Transport* London: Grosvenor Press International, 287–90.

Dunbar-Nobes, A.C. (1984) 'Port problems and small-island economies', in B.S. Hoyle and D. Hilling (eds) *Seaport Systems and Spatial Change*, Chichester: Wiley, 81–97.

Eastman, C.R. and Pickering, D. (1981) *Transport Problems of the Urban Poor in Kuala Lumpur,* Crowthorne, Transport and Road Research Laboratory, Report No 683.

Economic Commission for Africa (1975) *Study of Air Freight Potential in Developing Africa,* Addis Ababa, E/CN.14/TRANS/124.

Edmonds, G.A. (1980) 'The roads and labour programme, Mexico', in G.A. Edmonds and J.D.F.G. Howe (eds) *Roads and Resources,* London: Intermediate Technology Publications, 123–34.

Edmonds, G.A. and Howe, J.D.F.G. (1980) (eds) *Roads and Resources,* London: Intermediate Technology Publications.

Edwards, C. (1978) 'Some problems in evaluating investments in rural transport', Proceedings of a Transport Research Conference, Warwick: The University Press.

Eliot Hurst, M.E. (1973) 'Transportation and the societal framework', *Economic Geography*, 49: 163–80.

Ellis, C.I. (1979) *Pavement Engineering in Developing Countries,* Crowthorne: Transport and Road Research Laboratory, Report No SR 537.

Elmendorf, M. and Merrill, D. (1977) *Socio-economic Impact of Development in Chan Kom, 1971–6,* Washington: The World Bank, Development Economics Department.

ESCAP (1979) *Planning the Development of Inland Waterways,* Bangkok: Economic and Social Commission for Asia and the Pacific, 2 vols.

—— (1982) *Report of the Seminar-cum-Study Tour of Inland Waterway Terminals and Landings in China,* Bangkok: Economic and Social Commission for Asia and the Pacific.

—— (1984) *Developments in Shipping, Ports and Inland Waterways,* Bangkok: Economic Commission for Asia and the Pacific.

Etherington, K. (1993) *The Role of Cyclos in Urban Employment in Phnom Penh,* unpublished final year dissertation, Department of Geography, Royal Holloway, University of London.

Farahmand-Razavi, A. (1994) 'The role of international consultants in developing countries', *Transport Policy*, 1(2): 117–23.

Farrington, J. (1992) 'Transport, environment and energy', in B.S. Hoyle and R. Knowles *Modern Transport Geography*, London: Belhaven, 51–66.

325

Faulks, R. (1990) 'Bus provision in developing countries', in M. Heraty (ed.) *Developing World Transport*, London: Grosvenor Press International, 94–7.

Fernandez, J.E. and De Cea, J. (1991) 'An evaluation of the effects of deregulation policies on the Santiago, Chile, public transport system', in M. Heraty (ed.) op. cit., 127–33.

Field, H. (1989) 'Commercial aircraft for the developing world', in M. Heraty (ed.) *Developing World Transport*, London: Grosvenor Press International, 306–8.

Fieler, G. and Goodovitch, T. (1994) 'Decline and growth, privatisation and protectionism: the Middle East airline industry', *Journal of Transport Geography*, 2(1): 55–64.

Figueroa, O. and Henry, E. (1991) 'Analysis of the underground systems in Latin America', in M. Heraty *Developing World Transport*, London: Grosvenor Press International, 232–40.

Filani, M.O. (1993) 'Transport and rural development in Nigeria', *Journal of Transport Geography*, 1(4): 248–54.

Fishlow, A. (1965) *American Railroads and the Transportation of the Ante-Bellum Economy*, Cambridge, MA: Harvard University Press.

Flores, R.S. (1992) 'Metro Manila rediscovers the Pasig River', Venice: *Aquapolis*, 1(6): 32–7.

Fogel, R.W. (1962) *Railroads and American Economic Growth*, Baltimore: Johns Hopkins University Press.

Fong Chan Onn (1984) 'Johore port; its role in the growth of south peninsula Malaysia', Tokyo: *The Developing Economies*, 22(2): 185–204.

Fouracre, P.R., Maunder, D.A.C., Pathak, M.G. and Rao, C.H. (1981) *Studies of Bus Operations in Delhi, India*, Crowthorne: Transport and Road Research Laboratory, Report No SR 710.

Fouracre, P.R., and Maunder, D.A.C (1986) *A Comparison of Public Transport in Three Medium Sized Cities of India*, Crowthorne: Transport and Road Research Laboratory, Report No 82.

—— (1987) *Travel Demand Characteristics in Three Medium Sized Indian Cities*, Crowthorne: Transport and Road Research Laboratory, Report No 121.

Fowler, D. (1979) 'Financial analysis of railway operations', *Transport*, May: 281–8.

Frankel, E.G. (1986) 'Shipping and its role in economic development', *Marine Policy*, 13: 22–38.

—— (1987) *The World Shipping Industry*, London: Croom Helm.

Gallagher, R. (1992) *The Rickshaws of Bangladesh*, Dhaka: Dhaka University Press.

Gananathan, V.S. (1973) 'Roads as a means of surface transport in India', *National Geographic* (Allahabad), 8: 15–24.

Gardiner, R. (ed.) (1992) *The Shipping Revolution*, London: Conway Maritime.

Garrison, W.L., Berry, B.J.L., Marble, D.F., Morrill, R. and Nystuen, J. (1959) *Studies In Highway Development and Geographic Change*, Seattle: University of Washington.

Garrison, W.L. and Marble, D. (1962) *The Structure of Transportation Networks*, Evanston, IL: Northwestern University.

Gauthier, H. (1970) 'Geography, transportation and regional development', *Economic Geography*, 46: 612–9.

Gidwitz, B. (1980) *The Politics of International Air Transport*, Lexington, MA: D.C. Heath.

Gilman, S. (1991) *Asian container ship networks and hub port strategies*, Liverpool: The University, mimeo.

Glaister, G. (1980) 'The self-help approach: Afghanistan', in G.A. Edmonds and J.D.F.G.Howe (eds) *Roads and Resources*, London: Intermediate Technology Publications, 135–55.

Gold, E. (1981) *Maritime Transport*, Lexington, MA: D.C. Heath.

Goodland, R.J.A. and Irwin, H.S. (1975) *Amazon Jungle: Green Hell to Red Desert*, Amsterdam: Elsevier.

Gould, P.R. (1960) *Transportation in Ghana*, Evanston, IL: Northwestern University Studies in Geography, No 5.

Greenhill, B. (1971) *Boats and Boatmen in Pakistan*, Newton Abbot: David and Charles.

Greer, G. (1984) 'The Sao Francisco', in BBC, *River Journeys*, London: British Broadcasting Authority.

Hadfield, C. (1986) *World Canals: Inland Navigation Past and Present*, Newton Abbot: David and Charles.

Haggett, P. and Chorley, R. (1969) *Network Analysis in Geography*, London: Arnold.

Hall, S., Zegras, C., and Rojas, H.M. (1994) 'Transportation and energy in Santiago, Chile', *Transport Policy*, 1(4): 233–43.

Hammond, R. (1968) *Modern Methods of Railway Operation*, London: Muller.

Harris, V.A.P. (1989) 'Road maintenance – today's crisis, tomorrow's challenge', in M. Heraty (ed.) *Developing World Transport*, London, Grosvenor Press International, 31–6.

Harrison Church, R.J. (1956) 'The pattern of transport in British West Africa', in R.W. Steel and C.A. Fisher (eds) *Geographical Essays on British Tropical Lands*, London: Philips, 56–78.

Hathway, G. (1985) *Low Cost Vehicles*, London: Intermediate Technology Publications.

Hay, A. (1973a) *Transport for the Space Economy*, London: Macmillan.

—— (1973b) 'The importance of passenger transport in Nigeria', in B.S. Hoyle (ed.) *Transport and Development*, London, Macmillan, 125–38.

Hay, W.W. (1961) *Introduction to Traffic Engineering*, Chichester: Wiley.

—— (1982) *Railroad Engineering*, New York: Wiley.

Hayman-Joyce, J.G. (1979) 'River ports – Puerto Busch and Pucallpe', *The Dock and Harbour Authority*, 59: 374–83.

Hayuth, Y. (1981) 'Containerisation and the load centre concept', *Economic Geography*, 57: 160–88.

—— (1983) 'The evolution and competitiveness of air cargo transportation: the case of Israel's airborne trade', *Transport Reviews*, 3(3): 265–86.

—— (1987) *Intermodality*, London: Lloyd's of London Press.

Hepworth, M. and Ducatel, K. (1992) *Transportation in the Information Age*, London: Belhaven.

Heraty, M. (1980) *Public Transport in Jamaica*, Crowthorne: Transport and Road Research Laboratory, Report No SR 546.

—— (ed.) (1991a) *Urban Transport in Developing Countries*, London: PTRC.

—— (ed.) (1991b) *Developing World Transport*, London: Grosvenor Press International.

Heyerdahl, T. (1950) *The Kon Tiki Expedition*, London: George Allen & Unwin.

—— (1971) *The Ra Expedition*, London: George Allen & Unwin.

Hilling, D. (1966) 'Tema – the geography of a new port', *Geography*, 51(2): 11–25.

—— (1969a) 'The evolution of the major ports of West Africa', *Geographical Journal*, 135(3): 365–78.

—— (1969b) 'Saharan iron ore oasis', *Geographical Magazine*, 41(12): 908–17.

—— (1973) 'Container potential of West African ports', in B.S. Hoyle (ed.) *Transport and Development*, London: Macmillan, 151–66.

—— (1976a) 'Transport and the development process', *Proceedings of the British Association for the Advancement of Science*, E.85.

—— (1976b) 'Port congestion – the West African experience', *The Dock and Harbour Authority* December, 278–81.

—— (1977a) *Barge Carrier Systems – Inventory and Prospects*, London: Benn.

—— (1977b) 'The evolution of a port system – the case of Ghana', *Geography*, 62(2): 97–105.

—— (1977c) 'Lakeland route through Ghana', *Geographical Magazine*, 49(5): 308–12.

—— (1978) 'The infrastructure gap', in A. B. Mountjoy (ed.) *Third World – Problems and Perspectives*, London: Macmillan, 84–92.

—— (1980a) 'Water power – water-borne transport in West Africa', *West Africa*, 5 May: 791–4.

—— (1980b) 'Technological choice and maritime transport', *Future of Maritime Transport*, Cairo: Al Ahram, 17–22.

—— (1983) 'Ships, ports and developing countries', *Geoforum*, 14(3): 333–40.

—— (1985) 'Port efficiency and capacity: identifying the bottlenecks', *Proceedings Asia Port Management Conference*, Singapore: Portec.

—— (1990) 'African development and the maritime sector', *The Dock and Harbour Authority*, 70, April: 330–5.

Hoare, A.G. (1974) 'International airports as growth poles: a case study of Heathrow Airport', *Transactions Institute of British Geographers*, 63: 75–96.

Hodges, J.W., Rolt, J. and Jones, T.E. (1975) *The Kenya Road Transport Cost Study*, Crowthorne: Transport and Road Research Laboratory, Report No 673.

Hofmeier, R. (1972) *Transport and Economic Development in Tanzania*, Munchen: Weltforum Verlag.

Hofton, A. (1989) 'Developing world airlines – how they can survive and prosper', in M. Heraty (ed.) *Developing World Transport*, London: Grosvenor Press International, 309–12.

Holsman, A.J. and Crawford, S.A. (1974) 'The role of air transport in underdeveloped regions – North West Australia', *Journal Tropical Geography*, 39: 34–42.

Holt, C.D. (1989) 'The economics of traction policy for Developing Countries', in M. Heraty (ed.) *Developing World Transport*, London: Grosvenor Press International, 216–20.

Hoover, E.M. (1948) *The Location of Economic Activity*, New York: McGraw Hill.

Hornell, J. (1946) *Water Transport: Origins and Early Evolution*, Cambridge: Cambridge University Press.

Horridge, A. (1986) *Sailing Craft of Indonesia*, Oxford: Oxford University Press.

Howe, C. (ed.) (1969) *Inland Water Transportation: Studies in Public and Private Management and Investment Decisions*, Baltimore: Resources for the Future.

Howe, J. (1984) 'The impact of rural roads on poverty alllleviation – a review of the literature', in J. Howe and R. Richards (eds) *Rural Roads and Poverty Alleviation*, London: Intermediate Technology Publications, 48–81.

Howe, J. and Richards, P. (eds) (1984) *Rural Roads and Poverty Alleviation*, London: Intermediate Technology Publications.

Hoyle, B.S. (1973) 'Transport and economic growth in developing countries – the case of East Africa', in B.S. Hoyle (ed.) *Transport and Development*, London: Macmillan, 50–62.

—— (1983) *Seaports and Development: the Experience of Kenya and Tanzania*, London: Gordon and Breach.

—— (1988) *Transport and Development in Tropical Africa*, London: Murray.

—— (1992) 'Cities on water: the role of water transport in urban development', Venice: *Aquapolis*, 1(6): 8–13.

Hoyle, B.S. and Charlier, J. (1995) 'Inter-port competition in developing countries', *Journal of Transport Geography*, 3(2): 87–103.

Hoyle, B.S. and Hilling, D. (eds) (1970) *Seaports and Development in Tropical Africa*, London: Macmillan.

—— (eds) (1984) *Seaport Systems and Spatial Change*, Chichester: Wiley.

Hoyle, B.S. and Knowles, R. (eds) (1992) *Modern Transport Geography*, London: Belhaven.

Hoyle, B.S. and Pinder, D. (eds) (1981) *Cityport Industrialisation and Regional Development*, Oxford, Pergamon.

Hoyle, B.S. and Smith, J. (1992) 'Transport and development', in B.S. Hoyle and R. Knowles (eds) *Modern Transport Geography*, London: Belhaven, 9–31.

Hubbard, R. (1974) 'The development of the Jamaican road network by 1846', *Journal of Tropical Geography*, 38: 31–6.

Hughes, C. (1977) 'Containerisation in an LDC environment', *Maritime Policy and Management*, 4(5): 293–303.

Hughes, David E. (1981) *The Indonesian Cargo Sailing Vessels and the Problems of Technological Choice in Sea Transport*, unpublished Ph.D. thesis, University of Wales.

Hugill, S. (1967) *Sailortown*, London: Routledge & Kegan Paul.

Imakita, J. (1978) *A Techno-economic Analysis of Port Transport*, Farnborough: Saxon House.

Indo-American Chamber of Commerce (1975) *Barge Carrying Ships – their Feasibility to India's Seaborne Trade*, Bombay.

Institution of Civil Engineers (1986) *Moving People in Tomorrow's World*, London: Thomas Telford.

Jacobs, G.D. and Sawyer, I.A. (1983) *Road Accidents in Developing Countries*, Crowthorne: Transport and Road Research Laboratory, Report No SR 807.

Jacobs, G.D., Maunder, D.A.C. and Fouracre, P.R. (1986) 'A review of public transport operations in Third World cities', in Institution of Civil Engineers, *Moving People in Tomorrow's World*, London: Thomas Telford, 7–18.

Jansen, E.G., Dolman, A.J., Jerve, A.M. and Nazibor, R. (1989) *The Country Boats of Bangladesh*, Dhaka: University Press with Intermediate Technology Publications, London.

Janssen, J.O. and Shneerson, D. (1982) *Port Economics*, London: MIT Press.

—— (1987) *Liner Shipping Economics*, London: Chapman & Hall.

Jefferson, M. (1928) 'The civilising rails', *Economic Geography*, 4: 217–31.

Jenks, L.H. (1944) 'Railroads as an economic force in American development', *Journal of Economic History*, IV, 1.

Jin Goo Jim (1993) South Korean shipping, unpublished MSc dissertation, Marine Policy Programme, London School of Economics.

Jones, T. Hugh (1984) 'Rural roads and poverty alleviation in Thailand' in J. Howe and P. Richards (eds) *Rural Roads and Poverty Alleviation*, London: Intermediate Technology Publications, 142–62.

Joshi, P. (1981) The problems of urban transport, the case of Bombay, unpublished MPhil thesis, University of London.

Joy, S. (1971) 'Pricing and investment in railway freight services', *Journal of Transport Economics and Policy*, 5: 231–46.

Kaberry, N. (1989) 'Planning and designing an airport', in M. Heraty (ed.) *Developing World Transport*, London: Grosvenor Press International, 319–21.

Kansky, K.T. (1963) *Structure of Transport Networks: Relationships Between Network Geometry and Regional Characteristics*, University of Chicago: Department of Geography, Research Paper No 84.

Khan, M. (1987) 'A feasible inland navigation route from Kalabagh to Port Qasim, Pakistan', CIT Pakistan, *Proceedings of Seminar on Inland Water Transport*, Chartered Institute. of Transport Pakistan, Karachi, mimeo.

Klein, Martin S. (1966) 'The Atlantic highway in Guatemala', in G.W. Wilson *et al.*, *The Impact of Highway Investment on Development*, Westport, CT: Greenwood Press.

Kleinpenning, J.M.G. (1971) 'Road building and agricultural colonisation in the Amazon basin', *Tijdschrift voor Economische en Sociale Geografie*, 62: 285–9.

—— (1978) 'Further evaluation of the policy for the integration of the Amazon region, 1974–6', *Tijdschrift voor Economische en Sociale Geografie*, 69(1/2): 78–85.

Koenen, B.H.N. (1969) *The LAMCO Railroad*, Liberia, LAMCO JV, Railroad Department.

330

Kolars, J. and Malin, H.J. (1970) 'Population and accessibility: an analysis of Turkish railroads', *Geographical Review*, LX(2): 229–46.

Kosasih, D., Robinson, R. and Snell, J. (1987) *A Review of Some Recent Geometric Road Standards and their Application to Developing Countries*, Crowthorne: Transport and Road Research Laboratory, Research Report No 114.

Lachene, R. (1965) 'Networks and the location of economic activities', *Papers of the Regional Science Association*, 14: 183–96.

Lam, Han (1992) Minibus transport in Far Eastern cities with special reference to Beijing, unpublished Ph.D. thesis, University of London.

Lederer, A. (1979) 'Considerations a propos de la navigation sur les petite rivières tropicale', Brussels: *Bulletin Permanent International Association of Navigation Congresses*, 32: 11–18.

Lee, E.S.W. (1990) 'Formalising informal transport: paratransit in Hong Kong', in M. Heraty (ed.) *Developing World Transport*, London: Grosvenor Press International, 112–6.

Le Fevre, J. (1981) 'Urban buses: wind of change blows at home and abroad', *Transport*, March/April: 15–19.

Leinbach, T.R. (1975) 'Transport and the development of Malaya', *Annals Association American Geographers*, 65(2): 270–82.

—— (1976) 'The impact of accessibility upon modernising behaviour', *Tijdschrift voor Economische en Sociale Geografie*, 67(5): 279–88.

Leung, C.K. (1980) *China: Railway Patterns and National Goals*, University of Chicago: Department of Geography, Research Paper No 195.

—— (1982) 'The analysis and interpretation of national transport networks with special reference to the case of China', *Third World Planning Review*, 4(2): 177–91.

Levine, M.E. (1993) 'Scope and limits of multilateral approaches to international air transport', *International Air Transport*, Paris: OECD.

Lowe, J. and Moryadas, S. (1975) *The Geography of Movement*, Boston: Houghton Mifflin.

Lugard, F.D. (1922) *The Dual Mandate in British Tropical Africa*, Edinburgh: Blackwood.

Mabogunje, A.L. (1980) *The Development Process: a Spatial Perspective*, London: Hutchinson.

McCall, M.K. (1977) 'Political economy and rural transport: a reappraisal of transport impacts', *Antipode*, 9, 56–67.

McConville, J. (1995) *United Kingdom Seafarers*, London: Marine Society with Guildhall University.

McMaster, D. (ed.) (1970a) *Transport in Africa*, Edinburgh: Centre for African Studies, University of Edinburgh.

—— (1970b) 'Road communications and pattern of rural settlement', in D. McMaster (ed.) *Transport in Africa*, Edinburgh: Centre for African Studies, University of Edinburgh.

Maitland, D. (1981) *Setting Sails*, Hong Kong: South China Morning Post.

Majumdar, J. (1985) *The Economics of Railway Traction*, London: Gower.

Marc, R.C. (1973) 'La construction des routes par stades', *Industries et Travaux d'Outremer*, April, 246–50.

Marti, B. (1988) 'The evolution of Pacific Basin load-centres', *Maritime Policy and Management*, 15: 57–66.

Maunder, D.A.C. (1983) *Household and Travel Characteristics in Two Suburban Residential Districts of Delhi, 1982*, Crowthorne: Transport and Road Research Laboratory, Report No 767.

Maunder, D.A.C. and Fouracre, P.R. (1989) *Non-motorised Travel in Third World Cities*, Crowthorne: Transport and Road Research Laboratory, conference paper, mimeo.

Maunder, D.A.C., Fouracre, P.R., Pathak, M.G. and Rao, C.H. (1981) *Characteristics of Public Transport in Indian Cities*, Crowthorne: Transport and Road Research Laboratory, Report No 706.

Meinig, D.W. (1962) 'A comparative historical geography of two railnets: Columbia Basin and South Australia', *Annals Association of American Geographers*, 52(4): 394–413.

Miller, F. (1973) 'Highway improvements and agricultural production: an Argentine case study', in B.S. Hoyle (ed.) *Transport and Development*, London: Macmillan, 104–24.

Min. of Comm. (1985) *Chinese Highway and Water Transport*, Beijing: Ministry of Communications.

Mir, H.S. (1987) 'Energy conservation aspect of inland water transport', in CIT Pakistan *Proceedings of Seminar on Inland Water Transport*, Chartered Institute of Transport, Pakistan, Karachi, mimeo.

Nagorski, B. (1971) *Port Problems of Developing Countries*, Tokyo: International Association of Ports and Harbours.

Nash, C. (1985) 'European railway comparisons – what can we learn?', in K.J. Button and D.E. Pitfield (eds) *International Railway Economics*, Aldershot: Gower, 237–69.

Naveau, J. (1989) *International Air Transport in a Changing World*, London: Nijhoff.

N'Diaye, S. (1988) 'African airlines must cooperate to survive', *African Business*, July: 23–4.

Nock, N.S. (ed.) (1966) *Single Line Railways*, Newton Abbot: David and Charles for United Nations Economic Commission for Asia and the Far East.

O'Brien, P. (1983) *The New Economic History of Railways*, London: Croom Helm.

—— (ed.) (1977) *Railways and the Economic History of Western Europe, 1830–1914*, London: Macmillan.

Ocampo, R.B. (1982) *Low Cost Vehicles in Asia*, Ottawa: International Development Research Centre.

Ochia, K. (1990) 'Non-route travel in urban transportation planning', *Cities*, August: 230–5.

O'Connor, A.M. (1965) 'New railway construction and the pattern of economic development in Uganda', *Transactions Institute British Geographers*, 36: 21–30.

—— (1983) *The African City*, London: Hutchinson.

O'Flaherty, C.A. (1988) *Highways*, London: Arnold, 2 vols.

Ogundana, B. (1972) 'Oscillating seaport location in Nigeria', *Annals Association of American Geographers*, 62: 110–21.

—— (1973) 'Problems and prospects of river transport in Nigeria', Ibadan: *Nigerian Journal of Economic and Social Studies*, 15(3): 375–90.

Organisation for Economic Cooperation and Development (OECD) *Maritime Transport*, annual reports, Paris.

Owen,W. (1964) *Strategy for Mobility*, Washington: The Brookings Institution.

—— (1987) *Transportation and World Development*, London: Hutchinson.

Paquette, R.J., Ashford, N.J. and Wright, P.H. (1982) *Transportation Engineering*, Chichester: Wiley.

Patankar, P.G. (1986) 'Bombay's traffic problem', *Transport Reviews*, 6(3): 287–302.

Pawson, E. (1979) 'Transport and development: perspectives from historical geography', *International Journal of Transport Economics*, VI (2): 125–37.

Perrusset, A-C. (1977) 'Aménagement routiers en zone equatoriale forestière et accidenté', *Cahiers d'Outremer*, 30, 120: 404–11.

Persaud, W.H. (1986) *Air Jamaica: Contribution to the Jamaican Economy*, Mona: University of the West Indies, Institute of Social and Economic Research.

PIANC (1990) *Standardisation of Inland Waterways Dimensions*, Brussels: Permanent International Association of Navigation Congresses.

—— (1991) *Analysis of Cost of Operating Vessels on Inland Waterways*, Brussels: Permanent International Association of Navigation Congresses.

—— (1992) *Container Transport with Inland Vessels*, Brussels: Permanent International Association of Navigation Congresses.

Pirie, G.H. (1982) 'The de-civilizing rails: railways and underdevelopment in Southern Africa', *Tijdschrift voor Economische en Sociale Geografie*, 73(4): 221–8.

—— (1993) 'Transport, food insecurity and food aid in sub-Saharan Africa', *Journal of Transport Geography*, 1(1): 12–19.

Poernomosidhe, I.F. (1992) The impact of paratransit on urban road performance in the Third World, unpublished Ph.D. thesis, University of Wales.

Polak, G. and Koshal, P.R. (1980) 'Production function and changing technology for water transport', *Transportation Research*, 14a: 279–84.

Poole, M. (1990) 'Airport planning and operation', in M. Heraty (ed.) *Developing World Transport*, London: Grosvenor Press International, 248–52.

Porter, G. (1995) 'The impact of road construction on women's trade in rural Nigeria', *Journal of Transport Geography*, 3(1): 3–14.

Qiyu, W. (1987) 'The development of water transport in China', *Transport Reviews*, 7(1): 1–15.

Quastler, I.E. (1978) 'A descriptive model of railway network growth in the American Mid-West, 1865–1915', *Journal of Geography*, March: 87–93.

Ramamurti, R. (1985) 'High technology exports by state enterprises in LDCs: the Brazilian aircraft industry', Tokyo: *The Developing Economies*, 23(3): 254–80.

Reichmnan, S. (1965) 'Domestic air transport in Sierra Leone', *Bulletin Sierra Leone Geographical Association*, 9: 27–33.

Resende, E. (1973) *Rodovias e o Desenvolvimento do Brasil*, Rio de Janeiro, Ministerio dos Transportes.

Ridley, T.M. (1986) 'Metropolitan railways', in Institution of Civil Engineers, *Moving People in Tomorrow's World*, London: Thomas Telford, 19–49.

Rimmer, P.J. (1977) 'A conceptual framework for examining urban and regional transport needs in south-east Asia', *Pacific Viewpoint*, 18: 133–47.

—— (1982) 'Theories and techniques in Third World settings: trishaw pedallers and towkays in Georgetown, Malaysia', *Australian Geographer*, 15(3): 147–59.

—— (1984) 'The role of paratransit in South East Asian cities', *Singapore Journal of Tropical Geography*, 5(1): 45–62.

Rizvi, Z.H. (1987) 'Inland water transport in Pakistan – past, present and future', in CIT Pakistan, *Proceedings of Seminar on Inland Water Transport*, Chartered Institute of Transport Pakistan, Karachi, mimeo.

Robinson, R. (1976) 'Modelling the port as an operational system', *Economic Geography*, 52: 71–86.

—— (1989) 'The foreign buck: aid reliant investment strategies in Asean port development', *Transportation Research*, 23a(6): 439–51.

Robinson, R. (1981) *The Selection of Geometric Design Standards for Rural Roads in Developing Countries*, Crowthorne: Transport and Road Research Laboratory, Report No SR 670.

Rongju, T. (1984) 'The development of China's railway building', *Transport Reviews*, 4(1): 27–42.

Rostow, W.W. (1960) *The Stages of Economic Growth*, Cambridge: Cambridge University Press.

Rutz, W. (1987) 'Indonesia's sea transport – a series of maps', *Geo-Journal*, 14(4): 491–502.

Saggar, R.K. (1979) 'Comparative economics of inland water transport: a case study of the Ulhas River-Thana creek waterways, Bombay', in ESCAP, *Planning the Development of Inland Waterways*, Bangkok: ESCAP.

Sandberg, L. (1993) 'Project identification for inland transport activities on the lower Mekong', Brussels: *Bulletin International Association of Navigation Congresses*, 81: 20–8.

Sas, F.M., Blaauw, H.G. and Hollander, S. (1985) *The Advantages of Using Inland Waterways for Transportation*, Delft: Hydraulics Laboratory.

Savage, C.I. (1959) *An Economic History of Transport*, London: Hutchinson.

Schaeffer, K.H. and Sclar, E. (1975) *Access for All*, Harmondsworth: Penguin.

Schivelbusch, W. (1986) *The Railway Journey: the Industrialisation of Time and Space in the 19th Century*, Leamington: Berg.

Schumpeter, J.A. (1934) *The Theory of Economic Development*, Cambridge, MA: Harvard University Press..

Sessay, S.M. (1968) 'Factors influencing the pattern of road density in Sierra Leone', *Sierra Leone Geographical Journal*, 12: 17–26.

Shand, R.T. (ed.) (1980) *The Island States of the Pacific and Indian Oceans*,

Canberra: Australian National University, Development Studies Centre, monograph No 23.

Shawcross, W. (1984) 'The Mekong', in BBC, *River Journeys*, British Broadcasting Authority, 77–112.

Shneerson, D. (1981) 'Investment in port systems', *Journal of Transport Economics and Policy*, Sept.: 201–216.

Siddall, W.R. (1969) 'Railroad gauges and spatial interaction', *Geographical Review*, 59(1): 29–57.

Sjoberg, G. (1960) *The Pre-Industrial City*, New York: Free Press.

Slater, D. (1975) 'Underdevelopment and spatial inequality', *Progress in Planning*, 4: 99–67.

Smith, C. (1989) 'Planning and operating an international shipping service', in M. Heraty (ed.) *Developing World Transport*, London: Grosvenor Press International, 283–6.

Smith, J. (1970) *Road transport in Uganda*, Occasional Papers No 18, Department of Geography, University of Uganda.

Smith, J.A. (1989) 'Transport and marketing of horticultural crops by commercial farmers in Harare', *Geographical Journal of Zimbabwe*, 20: 1–15.

Smith, N.J. (1976) *Transamazon Highway*, Ann Arbor, Xerox University Microfilms.

Somabha, S. (1979) 'Mekong River', in ESCAP, *Planning the Development of Inland Waterways*, Bangkok: ESCAP.

Stanley, W.R. (1965) 'Air transport in Liberia: the role of the scheduled airline in the development of the Liberian transportation network', *Bulletin Sierra Leone Geographical Association*, 9: 34–44.

—— (1984) 'Changing export patterns of Liberia's seaports', in B.S. Hoyle and D. Hilling (eds) *Seaport Systems and Spatial Change*, Chichester: Wiley, 435–59.

Stokes, C.J. (1968) *Transportation and Economic Development in Latin America*, New York: Praeger.

Stopford, M. (1988) *Maritime Economics*, London: Unwin Hyman.

Taaffe, E.J. and Gauthier, N.L. (1973) *Geography of Transportation*, Englewood Cliffs, NJ: Prentice-Hall.

Taaffe, E.J., Morrill, R.L. and Gould, P.R. (1963) 'Transport expansion in underdeveloped countries: a comparative analysis', *Geographical Review*, 53: 503–29.

Takel, R. (1974) *Industrial Port Development* Bristol: Scientechnica.

Taneja, N.K. (1989) *Introduction to Civil Aviation*, Lexington, MA: D.C. Heath.

Taylor, Z. (1984) 'Seaport development and the role of the state: the case of Poland', in B.S. Hoyle and D. Hilling (eds) *Seaport Systems and Spatial Change*, Chichester: Wiley, 217–38.

Thomas, P.K. (1984) 'Rural roads and poverty alleviation in India', in J. Howe and P. Richards (eds) *Rural Roads and Poverty Alleviation*, London: Intermediate Technology Publications.

Thomson, J. Michael (1977) *Great Cities and Their Traffic*, Harmondsworth: Penguin.

Tolofari, S.R. (1984) 'Nigerian waterways – development problems and potential', *The Dock and Harbour Authority*, 64: 185–9.

—— (1989) *Open Registry Shipping*, London: Gordon and Breach.

Turner, D.L. (1991) *Container Shipping in the ESCAP Region*, paper presented at the 4th World Container and Intermodal Conference, Kuala Lumpur.

Ullman, E.L. (1974) 'Geography as spatial interaction', in M.E. Eliot Hurst (ed.) *Transportation Geography*, New York: McGraw Hill, 29–40.

UNCTAD (1973) *Berth Throughput – Systematic Methods for Improving General Cargo Operations*, New York: United Nations.

—— (1976) *Port Congestion: Report of the Group of Experts*, Geneva: United Nations Conference on Trade and Development, TD/B/C.4/152.

—— (1980) *The Rationalisation of Port Surcharges*, Geneva: United Nations Conference on Trade and Development, TD/B/C.4/202.

—— (1985) *Port Development – a Handbook for Planners in Developing Countries*, New York: United Nations.

United Africa Company (1957) 'Port capacity and shipping turnround in West Africa', *Statistical and Economic Review*, No 19.

United Nations (1967) *Transport Development*, New York: Economic and Social Council.

—— (1994) *Review of Maritime Transport, 1993*, Geneva: United Nations, E.94.II.D.30.

Vance, J.E. (1986) *Capturing the Horizon: the Historical Geography of Transportation*, London: Harper & Row.

Van den Heuvel, H. (1993) 'Container transport by inland waterways', Brussels: *Bulletin Permanent International Association of Navigation Congresses*, 80: 24–8.

Ville, S.P. (1987) *English Shipowning During the Industrial Revolution*, Manchester: Manchester University Press.

Voigt, F. (1967) *The Importance of the Transport System for Economic Development Processes*, Addis Ababa: United Nations Economic Commission for Africa.

Vorobjev, P. (1989) 'World shipping and maritime trade', in M. Heraty (ed.) *Developing World Transport*, London: Grosvenor Press International, 275–82.

Wahab, I.B. (1990) 'Urban transport in Kuala Lumpur', *Cities*, August: 236–43.

Wallace, W.H. (1958) 'Railroad traffic densities and patterns', *Annals Association American Geographers*, 48: 352–74.

Walling, D.E. (1981) 'Yellow River which never runs clear', *Geographical Magazine*, 53(9): 568–75.

Walters, A.A. (1979) 'The benefits of minibuses: the case of Kuala Lumpur', *Journal Transport Economics and Policy*, 13: 320–34.

Walters, A.A. (1980) 'The benefits of minibuses', *Journal Transport Economics and Policy*, 15: 79–80.

Ward, M. (1970) *The Rigo Road*, Canberra: Australian National University, New Guinea Research Unit, Bulletin No 33.

Waters, II, W.G. (1985) 'Rail cost analysis', in K.J. Button and D.E. Pitfield (eds) *International Railway Economics*, Aldershot: Gower.

Wells, A.J. (1984) *Air Transportation*, Belmont, California: Wadsworth.

Wergeland, T. (1985) 'UNCTAD Liner Code', *Lloyd's Shipping Economist*, February: 6–9.

Wesche, R. (1974) 'Planned rainforest farming on Brazil's Transazamonica', *Revista Geografica*, (Mexico), 81: 105–14.

White, H.P. (1970) 'The morphological development of West African seaports', in B.S. Hoyle and D. Hilling (eds) *Seaports and Development in Tropical Africa*, London: Macmillan, 11–25.

White, N.P. and Senior, M.L. (1983) *Transport Geography*, Harlow: Longman.

Wiese, B. (1984) 'The role of seaports in the industrial decentralisation process: the case of South Africa', in B.S. Hoyle and D. Hilling (eds) *Seaport Systems and Spatial Change*, Chichester: Wiley.

Wilbanks, T. (1972) 'Accessibility and technological change in North India', *Annals Association of American Geographers*, 62(3): 427–36.

Williams, G.J. and Hayward, D.F. (1973) 'Recent development of the Sierra Leone Airways system', *Geography*, 58(1): 58–61.

Wilson, G.W. (1973) 'Towards a theory of transport and development', in B.S. Hoyle (ed.) *Transport and Development*, London: Macmillan, 208–30.

Wilson, G.W., Bergmann, B.R., Hirsch, L.V. and Klein, M.S. (1966) *The Impact of Highway Investment on Development*, Westport, CT: Greenwood Press.

Wood, J.P. (1990) 'Port design and construction', in M. Heraty (ed.) *Developing World Transport*, London: Grosvenor Press International, 213–7.

Wright, Charles L. (1992) *Fast Wheels Slow Traffic*, Philadelphia: Temple.

Yeats, A.J. (1981) *Shipping and Development Policy*, New York: Praeger.

Yue-Hong Chou, (1993) 'Airline deregulation and nodal accessibility', *Journal of Transport Geography* 1(1): 36–46.

Zhihao, W. (1990) 'Bicycles in large cities in China', in M. Heraty (ed.) *Developing World Transport*, London: Grosvenor Press International, 130–4.

Zuogao, W. (1990) 'Navigation on the Yangtze River and the Three Gorges project', Brussels: *Bulletin Permanent International Association of Navigation Congresses*, 70: 86–96.

# INDEX

Pacific Rim 115, 144, 260, 291
Pakistan: aerospace industry 153; ports
260; shipping 292; water transport
40, 69
Panama: shipping 284, 285–6, 304, 305
Papua New Guinea: air transport 121,
130, 132; roads 19, 169, 170, 182–4;
water transport 49
Paraguay: air transport 120, 121
Paraguay (river) 56, 66
Parana (river 40, 45, 46, 48, 51, 59, 61
passenger demand/transport 5, 19, 72,
111, 116–19, 157, 181, 192, 202,
203, 25–9, 218, 224, 226; see also
urban transport and vehicles
Patel On-Board Couriers 242
Penang 239
permissive impact 12
Peru: air transport 120, 121; ports 272,
273; shipping 295, 297
Philippines: aerospace industry 153; air
transport 120, 121, 129, 130; IPT
213, 217; ports 257, 260, 273 railway
109; shipping 276, 286, 294, 295,
298, 305, 306, 312; urban transport
209; vehicle assembly 217; water
transport 59, 72–3
Phnom Penh 225, 226, 227, 228
Phuket 239
Pichilinque 259
pilgrim travel 124
Ping Hai Wan 200
'pinisi' 277
Point Lisas 252
Ponta da Madeira 243, 250
Port Harcourt 259
Port Kelang 241, 273
'porpuestos' 213
positive impact 12
post-modernists 222
prior dynamism 10, 11, 13
privatisation 139
'publicos' 215, 218
public light bus (PLB) 213, 218,
219–20
public versus private ownership 316–18
Puerto Bolivar 250
Puerto Rico: IPT 218; shipping 295
Puerto Ventanas 273

Pune 226
push towing 48, 55ff

Quequen 273
Quintero Bay 273

railways and development 10, 12, 16,
76–9; economics 99–101; gauge
87–90; investment 104–10;
maintenance 101–2; networks
102–4; organisation 79–81; paths/
capacity 94–9; technology 81–4;
track 84–91; traction 91–4
Rand 26
Red Sea 59
Redi 245, 248
Rhine, the 43, 56, 58, 71
Richard's Bay 248
'rickshaws' 25, 224, 225
Rigo road 19, 182
Rio de Janiero 233
river navigation 41–52
river ports 62–6
roads and accessibility 171–6;
construction technology 164–9;
consequences of improvement
181–91; design and capacity 169–71;
earth roads 158, 161–3, 167; gravel
roads 163–4, 165, 166, 167, 169,
170, 175, 177, 178, 179, 186; low-
cost roads 176–9; motorways 24,
158, 174, 175, 189; networks
158–64; planned road improvement
179–81; rural access 170, 177–9,
180–1, 188–91 road development
schemes 185–91; safety 171;
seasonal availability 162, 164;
surfaced roads 164, 165, 167, 168,
175; vehicles 191–4
Rosario 273
route sufficiency 180–1
'rubber' cities 202–4

Sabah 17, 19
Sahel 124, 173
Saigon 200
Saldanha Bay 241, 248
Salvador 233
Salween, (river) 59

For Product Safety Concerns and Information please contact our EU
representative GPSR@taylorandfrancis.com Taylor & Francis Verlag GmbH,
Kaufingerstraße 24, 80331 München, Germany


Printed and bound by CPI Group (UK) Ltd, Croydon, CR0 4YY
08/05/2025
01864494-0002